Statins: Indications and Uses, Safety and Modes of Action

STATINS: INDICATIONS AND USES, SAFETY AND MODES OF ACTION

GUNNAR N. HOLMQVIST
EDITOR

Nova Science Publishers, Inc.
New York

For permission to use material from this book please contact us:
Telephone 631-231-7269; Fax 631-231-8175
Web Site: http://www.novapublishers.com

LIBRARY OF CONGRESS CATALOGING-IN-PUBLICATION DATA

Available upon request

ISBN: 978-1-60692-103-6

Published by Nova Science Publishers, Inc. † New York

Contents

PREFACE

This new book focuses on statins which are a relatively new group of drugs used to lower blood cholesterol levels. A high cholesterol level increases a person's risk of having a heart attack or stroke. The long-term use of statins reduces the risk of such an event and can increase the life expectancy of people with a history of heart disease. The statins work by blocking an enzyme in the body that is involved in the production of LDL cholesterol, especially in the liver. This enzyme is known as HMG coenzyme A reductase. The statins are the most effective group of drugs for lowering the levels of LDL cholesterol in the body. Potential side-effects include muscle cramps and gastrointestinal upsets. These are usually resolved on temporarily lowering the dose. Liver enzyme derangements may occur, which generally return to normal after briefly discontinuing the drug. Some report headaches. Other side-effects occur rarely.

Chapter 1 - The authors propose a new concept, 'Vascular Failure,' to detect early stage atherosclerosis, which is characterized as integration of endothelial dysfunction, smooth muscle cell dysfunction and metabolic abnormality of the vessel wall. Also, 'Vascular Failure' occurs not only in atherosclerosis, but also in vasculitis as well as systemic inflammatory disorders, which is of great interest. Statins targeting 'Vascular Failure' should be appled in the earlier stage, even when anatomical vascular abnormalities are not present.

Chapter 2 - Statins have been shown to be remarkably potent in reducing cardiovascular events and improving patient's survival. Overall, the statins show a rather good safety profile. Apart from reports of rare life threatening side effects, mainly from rhabdomyolysis (particularly in combination treatment with fibrates), knowledge and understanding on a great variety of different side effects occurring quite frequently is limited until now. This review on side effects in statin monotherapy attempts to analyse whether there are predisposing factors or specific properties of one or the other of the compounds concerning the development of side effects such as hepatic and renal function, erectile dysfunction, muscular problems and others. Types, prevalence and association with concurrent diseases are described. Underlying biochemical aspects, such as oxidation injury and eventual therapeutic interventions, are discussed. It seems that the rare severe statin side effects can completely be avoided by adhering to 5 simple rules, while the mild side effects probably are much more prevalent than considered so far.

Chapter 3 - The statins are widely used for the prevention of cardiovascular diseases thanks to their lipid lowering effects. Recent reports suggest that statins have antihypertensive and antiinflammatory properties besides the lipid lowering one. Moreover, in animal models of hypertension and kidney disease, the statins induced both the eNOS up-regulation and the iNOS, LOX-1 and NFkB down-regulations in the renal endothelium promoting the increase of the perfusion lowering the sclerosis as well. All these effects of statins might be helpful in reducing the incidence of cardiovascular diseases in human hypertensives whose clinical picture is often characterized by risk factor such as dyslipidemia, elevated serum C-reactive protein levels and nephrosclerosis with reduced renal vascular reserve and microalbuminuria.

Aim of this prospective, self-controlled, interventional study was to assess the trends of the renal hemodynamics, the blood pressure, the lipid panel and the inflammation during a short course of atorvastatin therapy, in essential hypertensives (EH).

After a 5-days run-in period, the patients began atorvastatin therapy (10 mg q.d.). At T_0 (before the first tablet of atorvastatin) and T_{30} (after 30 days of therapy) the patients underwent the following assessments: systolic, diastolic and mean blood pressure (SBP, DBP and MAP respectively), glomerular filtration rate (GFR) and effective renal plasma flow (ERPF) (inulin and p-aminohippurate clearance respectively), filtration fraction (FF) (GFR/ERPF) and total renal vascular resistances (TRVR) (using ERPF, hematocrit and MBP), microalbuminuria and serum levels of lipids and C-reactive protein (hs-CRP). The patients followed a standard diet (35 Kcal/kg bw/day, 1 g of protein/kg bw/day and 6 g of sodium/day). Never treated patients with I stage EH were included. Patients with heart, kidney or liver diseases were excluded as well as patients taking any other medication.

This report provides preliminary data about 6 EH (3M/ 3F), 45.5 ± 14.8 (mean ± SD) years old so far enrolled. The patients have been compliant with the atorvastatin therapy as showed by the reduction of serum lipids (mg/dL) (total cholesterol: T_0: 226.8 ± 33.7 vs T_{30}: 164.5 ± 46.5, p=0.01; triglycerides: T_0: 157.8 ± 50.2 vs T_{30}: 89.8 ± 16.5, p=0.05; apo-B: T_0: 110.3 ± 24.2 vs T_{30}: 75.5 ± 28.8, p=0.003). Mean daily protein and sodium intake as well as serum hs-CRP levels, microalbuminuria and renal hemodynamics were unchanged during the study instead of blood pressure which was significantly reduced (mmHg) (SBP: T_0 : 150.3 ± 10.6 vs T_{30}: 139.1 ± 6.9, p=0.01; DBP: T_0 : 97.4 ± 11.5 vs T_{30}: 88.6 ± 9.4, p=0.04; MAP: T_0 : 115.0 ± 11.0 vs T_{30}: 105.5 ± 7.9, p=0.02).

This preliminary experience on EH suggest that in EH a short course of atorvastatin therapy is associated with the statistically significant reduction of both the serum lipids levels and the blood pressure and the restoration of the glomerular autoregulation. The absence of changes of the serum hs-CRP levels and microalbuminuria during atorvastatin therapy might mean that the pleiotropic effects of statins demonstrated in animal models might be subdivided in short-term and long-term effects, depending on the type, the dose and the timing of therapy.

Chapter 4 - *Objective*: The objective of the present investigation was to find out whether the addition of fenofibrate to statin monotherapy produced any synergistic or additive beneficial effects in reducing risk factors, especially plasma fibrinogen, in patients of Acute Coronary Syndromes (ACS) requiring Percutaneous Coronary Interventions (PCI).

Methods: This was a randomized, non-blind, prospective study with parallel group design, conducted in 102 patients who had angiographically documented Coronary Artery Disease (CAD). All had undergone angioplasty. The patients were randomized to atorvastatin (20mg/day, n=25), simvastatin (40mg/day, n=27), atorvastatin(10mg/day)-fenofibrate (200mg/day) combination (n=25) and simvastatin (20mg/day)-fenofibrate(200mg/day) combination (n=25). The serum lipid profile and plasma fibrinogen were recorded before initiation of therapy and after 3 months of the respective treatments.

Results: All the patients already had desirable lipid levels as per the NCEP ATP III guidelines. The addition of fenofibrate to statin monotherapy produced additional benefits on reduction in triglyceride (TG) and very low density lipoprotein (VLDL) levels, and an increase in high density lipoprotein (HDL) levels. All the treatment groups showed a significant decrease in the plasma fibrinogen levels. This did not correlate with any of the study parameters like age, body weight, hemodynamic characteristics and lipoprotein levels. Statin monotherapy produced a significant decrease in the fibrinogen levels and the addition of fenofibrate further enhanced the reduction.

Conclusions: Addition of fenofibrate to statins seems to be beneficial in patients with ACS. Statins, contrary to various reports, were found to decrease plasma fibrinogen significantly.Furhter, in combination with fenofibrate there was enhanced reduction of the novel risk factor, fibrinogen.

Chapter 5 – The beneficial effect of statins on the reduction of cardiovascular events can be only partly attributed to their cholesterol lowering effect. The antiproliferative, anti-inflammatory, and immunomodulatory properties of statins appear to be largely unrelated to lipid-lowering but may be explained by affecting post-translational modification or isoprenylation essential for membrane localization and biologic activity of several proteins including adapter proteins and enzymes involved in signal transduction pathways. The present data reinforce the hypothesis that statins may represent innovative pharmacological tools not only for the prevention of cardiovascular related disease in normolipidemic patients but also for diseases where a reduced isoprenylation of regulatory proteins reveals benefit effects.

Chapter 6 - It has been repeatedly shown that statins (3-hydroxy-3-methylglutaryl coenzyme A reductase inhibitors) are very effective in primary and secondary prevention of ischemic heart disease. In the settings of acute coronary syndrome (ACS), different pathological pathways are triggered that are known to be inhibited by statins, including endothelial dysfunction and activation of inflammation and coagulation; the idea to use these drugs also under conditions of ACS seems to be, therefore, fully justified. Recently, several prospective controlled clinical trials have been presented, showing safety and in some points also efficacy of statins, when administered early after ACS. An increasing number of publications demonstrates, however, that statins may express a positive effect not only in the early secondary prevention but also directly in the therapy of ACS, i.e. when statin treatment is started as a first-line care in clinically unstable patients. This therapeutic option is supported by (i) experimental studies, showing a protective effect of statins under the condition of acute ischemia, (ii) analysis of different registers and trials, demonstrating better prognosis of statin-treated patients, and (iii) small clinical trials, describing a lower peri-procedural infarction rate during coronary intervention or lower level of C-reactive protein and other inflammatory markers as a result of statin therapy. Nevertheless, confirmation of this hypothesis by large prospective controlled clinical trials will be necessary before introduction of statins as the first line therapy in unstable patients with ACS, even without knowledge of the blood cholesterol level.

Chapter 7 - Lowering high levels of low-density lipoprotein cholesterol (LDL-C) is the primary aim in the prevention of cardiac events. However, low levels of high-density lipoprotein cholesterol (HDL-C) are also associated with an increased risk of ischemic heart disease.

Lipid-lowering drugs are known to decrease LDL-C and to increase HDL-C slightly. However, not all patients benefit from this effect. Some patients have lower HDL-C during statin treatment than before the treatment.

These patients were first described in a case report in 2002 as ‘bad HDL-C responders to statins’. In the case of one man, HDL-C and the ratios of total cholesterol (TC) to HDL-C and LDL-C to HDL-C worsened dramatically during pravastatin treatment. After 3 years, pravastatin was replaced by fenofibrate. The result was spectacular. The HDL-C increased to at least twice the level obtained during pravastatin.

Bad HDL-C responders are characterized by HDL-C levels which decrease below 40 mg/dl during the treatment, despite higher HDL-C levels before the treatment.

The existence of bad HDL-C responders to statins was confirmed by a prospective survey of 2,259 patients treated with a statin or a fibrate for hyperlipidaemia. The proportion of bad HDL-C responders is higher for statins (6%) than for fibrates (4%).

In a review of the guidelines, almost all selected guidelines consider low HDL-C as a marker of increased risk for coronary heart disease. However, only few guidelines use the level of HDL-C as a threshold or target level for the treatment of dyslipidemia. The guidelines provide only little information on the management of patients with treatment-induced low HDL-C. Instead of using TC or LDL-C we consider the use of the ratios of TC to HDL-C or LDL-C to HDL-C as a threshold as well as a target for treatment.

Treatment with fibrates was studied in 14 bad HDL-C responders to statins. Far better levels for HDL-C, TC to HDL-C and LDL-C to HDL-C were obtained with fibrates compared to statins. For bad HDL-C responders to statins with low or normal LDL-C, treatment with fibrates instead of statins should be considered. For those with high LDL-C, fibrates should be added to statins.

Treatment for bad HDL-C responders should be studied in randomized controlled trials. Such a trial with simvastatin and fenofibrate has been initiated to corroborate the findings.

Chapter 8 - The 3-hydroxy-3-methylglutaryl-coenzyme-A (HMG-CoA)-reductase inhibitors (statins) are the most commonly prescribed agents for the treatment of hypercholesterolemia, due to their efficacy in lowering LDL-cholesterol and ability to reduce clinical outcome in both primary and secondary prevention of coronary artery disease. In addition to their serum lipid-lowering action, statins display non lipid-lowering pharmacological activities known as pleotropic actions. The pleiotropic effects include significant anti-inflammatory and immunomodulatory actions and many essential cellular functions including cell proliferation, differentiation, and survival and participate in the regulation of cell shape and motility.

This chapter is about the intracellular pathways involved in the pathogenesis of systemic lupus erithematosus (SLE) and the possible effects of the statins in those pathways which could be modulating their therapeutic effects, including their beneficial effects on primary and secondary prevention of cardiovascular diseases, anti-inflammation, and immunomululation.

In: Statins: Indications and Uses, Safety and Modes of Action ISBN 978-1-60692-103-6
Editor: Gunnar N. Holmquist

Chapter 1

STATIN THERAPY FOR CORONARY ARTERY DISEASE BEYOND LIPID LOWERING EFFECT

Teruo Inoue and Koichi Node
Department of Cardiovascular and Renal Medicine, Saga University Faculty of Medicine, Saga, Japan

ABSTRACT

We propose a new concept, 'Vascular Failure,' to detect early stage atherosclerosis, which is characterized as integration of endothelial dysfunction, smooth muscle cell dysfunction and metabolic abnormality of the vessel wall. Also, 'Vascular Failure' occurs not only in atherosclerosis, but also in vasculitis as well as systemic inflammatory disorders, which is of great interest. Statins targeting 'Vascular Failure' should be appled in the earlier stage, even when anatomical vascular abnormalities are not present.

INTRODUCTION

3-hydroxy-3-methylglutaryl coenzyme A (HMG-CoA) reductase inhibitors, or statins, are es lished as the principal and the effective class of drug to reduce serum cholesterol levels. Although several different statins are available for treatment of hypercholesterolemia and their pharmacokinetic profiles are different, all statins have one characteristic in common. Statins inhibits the conversion of HMG-CoA to mevalonic acid with consecutive attenuation of the biosynthesis of cholesterol, reducing cellular cholesterol content in hepatocytes. Hepatocytes respond to sterol depletion by activating nuclear sterol regulatory element-binding protein-2, which upregulates the transcription of key genes implicated in cholesterol metabolism including HMG-CoA reductase and the low density lipoprotein (LDL) receptor. Thus, the cholesterol-lowering effect of statins is principally mediated by the up-regulation of

LDL-receptor activity, which leads to enhanced hepatic uptake of atherogenic apo B-containing lipoproteins, such as very low density lipoprotein (VLDL), VLDL remnant, intermediate density lipoprotein (IDL), and LDL.

There are increasing numbers of evidences that statins reduce cardiovascular events such as coronary artery disease in hypercholesterolemic patients in both primary and secondary prevention. The striking benefit achieved with statin treatments in patients with a wide range of cholesterol levels cannot be merely attributed to their cholesterol lowering effect. Recent substantial data has accumulated showing that statins exert various effects on multiple targets, namely pleiotropic effects, especially targeting blood vessels, which are considered to be derived from suppression of the small GTP-binding protein Rho and Rho kinase signaling, independently of cholesterol lowering properties. These effects include the improvement of vascular endothelial function, inhibiting vascular smooth muscle cell proliferation and migration, anti-inflammatory actions, anti-oxidative effects or stabilizing vulnerable plaques.

Atheroscrelosis is a progressive disease characterized as a response of the vessel wall to chronic, multifactorial injury and leads to the formation of atheromatous or fibrous plaques. These plaques are regions of thickened intima and are composed of various mixtures of fibrous tissues, cells, and lipid. Vessel wall injury promote endothelial dysfunction and increased adhesion of leukocytes and platelets to endothelium, leading to release of various inflammatory mediators that potentiate vascular smooth muscle proliferation, to accumulation of peroxidized lipid and thereby to plaque formation. In these processes, endothelial dysfunction is thought to be an initial stage of atherosclerosis. In addition to endothelial dysfunction, smooth muscle cell dysfunction, metabolic abnormality of the vessel wall including inflammation, oxidative stress, breakdown of neurohormonal balance occur in early stage of atherosclerosis process. Patients who have only risk factors as dyslipidemia as well as hypertension, diabetes mellitus, smoking or others are known to have endothelial dyfunction, smooth muscle dyfunction and abnormalities of the vessel wall metabolism. In light of early intervention in atherogenic risk factors to prevent disease progression, however, we believe these vascular functions should be integrated. Now, we propose a new concept 'Vascular Failure' to detect early stage atherosclerosis, which is characterized as integration of endothelial dysfunction, smooth muscle cell dysfunction and metabolic abnormality of the vessel wall. Also, 'Vascular Failure' not only in atherosclerosis but also in vasculitis or systemic inflammatory disorders is of great interest. We should try to apply the statins targeting 'Vascular Failure' in earlier stage even when anatomical vascular abnormalities are not present.

Prevention of Coronary Events by Statins in Clinical Trials

European WOSCOPS (West of Scotland Coronary Prevention Study) [1] and US AFCAPS/TexCAPS (Air Force/Texas Coronary Atherosclerosis Prevention Study) [2] are representative large clinical trials for the primary prevention of coronary events by statins. WOSCOPS is a randomized placebo-controlled trial in a double blind fashion that evaluated the effects of 40 mg/day of pravastatin on preventing the onset of coronary events in 6595 hyperlipidemic male subjects aged 45 to 65 years by the 4.9 year's observation. Although there were no significant differences in non-cardiac death, the onset of non-fatal myocardial infarction or death resulted from coronary artery disease was seen in 5.9 % in the pravastatin group that was significantly lower than 7.9 % in the placebo group, and the results showed 31 % (95 % CI; 17-45 %) of the relative risk reduction by pravastatin. In AFCAPS/TexCAPS [2], 20-40 mg/day of lovastatin or placebo was prescribed in a double blind fashion in total of 6605 subjects (5608 male aged 45-73 years and 997 menopaused females aged 55 to 73 years) including normolipidemic subjects. The results of 5.2 year's observation showed that the major coronary events appeared in 6.8 person/1000 person every year in lovastatin group and 10.9 person/1000 person every year in placebo group, indicating reduction of the coronary events by lovasatin with a 0.63 (95 % CI; 0.50-0.79) of the relative risk.

The clinical trials for secondary prevention of coronary artery disease includes 4S (Scandinavian Simvastatin Survival Study) [3], LIPID (Long-term Intervention with Pravastatin in Ischemic Disease Study) [4] and CARE (Cholesterol and Recurrent Events) [5]. The 4S evaluated the effects of simvastatin administration for 5.4 years in 4444 hypercholesterolemic subjects. As a result, in the simvastatin group, 34 %, 42 % and 39 % reductions of recurrent coronary artery disease, coronary death and total death, respectively, were evident in association with lipid lowering effects, compared with the control group. In the LIPID trial that evaluated the effects of pravastatin in 9014 subjects including nomolipidemic and mild to moderate hyperlipidemic subjects, recurrent coronary artery disease, coronary death and total death were also reduced. The CARE observed the effects of pravastatin in 4159 post myocardial infarction subjects with normal total cholesterol level and with normal to mild elevation of LDL cholesterol level. The 5 years' observation showed that pravastatin could reduce occurrence of non-fatal myocardial infarction or fatal coronary artery disease by 24 % and 27 %, respectively.

The AVERT (Atorvastatin Versus Revascularization Treatment) [6] is a moderate scale but very unique trial, in which aggressive lipid lowering therapy using 80 mg/day of atorvastatin was compared with percutaneous coronary intervention (PCI) with conventional medical treatments in 341 stable coronary artery disease subjects. The results of 18 months follow-up showed 36 % reduction of coronary events in atorvastatin group, compared with PCI group. In the MIRACL (Myocardial Ischemia Reduction with Aggressive Cholesterol Lowering) trial [7], in 3086 acute coronary syndrome (ACS) patients the initiation of aggressive lipid lowering by 80 mg/day of atorvastatin within 24-96 hrs from the onset reduced the ischemic events requiring readmission within 16 weeks from the onset. The LIPS (Lescol Intervention Prevention Study) [8] is a trial that compared the early initiation of 80 mg/day of fluvastatin with that of placebo in 1677 patients getting initial success of PCI. The results of 3 years' observation showed that 22 % reduction of the relative risk for major cardiac events in the fluvastatin group, compared with placebo group.

Statins' Lipid-Independent Effects for Vascular Failure

Statins' Actions for Vascular Endothelial Dysfunction

Endothelium is a flat monolayer of cells that cover vascular lumina throughout the body. Endothelial cells not merely are constitutes of the vessel wall, but play various biological roles, such as maintaining vascular tone and structure, regulating intravascular hemostasis and permeability, protecting from oxidative stress, and inhibiting cell adhesion and migration, i.e., anti-inflammatory properties [9]. Recent progress in vascular biology has revealed that the endothelium release a large number of vasactive substances. These substances are divided into two classes: endothelium-derived relaxing factors (EDRFs), and endothelium-derived contracting factors (EDCFs). It has been shown that EDRFs such as nitric oxide (NO), endothelium-derived hyperpolarizing factors (EDHF), or prostacyclin (PGI2) protect vasculature from atherogenic insult, whereas EDCFs such as endothelin-1 (ET-1) or thromboxane A2 (TXA2) have opposite effects and participate in the progression of cardiovascular diseases. Endothelial dysfunction is characterized by a reduction of the bioavailability of EDRFs, in particular, NO, whereas EDCFs increase [10]. This imbalance leads to an impairment of endothelium-dependent vasodilation, which represents the functional characteristics of endothelial dysfunction. On the other hand, endothelial dysfunction, aside from denoting impaired endothelium-dependent vasodilation, also comprises a specific state of "endothelial activation", which is characterized by a proinflammatory, proliferative, and procoagulatory milieu that favors all stages of atherogenesis [11]. Given this relationship between endothelial dysfunction and atherosclerosis, it is likely that the status of endothelial function may reflect the propensity of an individual to develop atherosclerotic disease, and thereby, the presence of endothelial dysfunction may serve as an indicator to detect the initial step of 'Vascular Failure'.

One of the best documented effects of statins for coronary vascular failure is improvement in parameters associated with endothelial dysfunction. Hypercholesterolemia reduces NO production and enhances its degradation in vascular endothelial cells. Statins' lipid-lowering effects improve these endothelial dysfunctions, increasing endothelial nitric oxide synthase (eNOS) gene expression via the reduction of degradation of mRNA and decreasing ET-1 production [12]. However, lipid-independent statins' effects on endothelial function may be more important, and such effects have been experimentally evident in the coronary artery. It has been recently demonstrated that simvastatin preserves coronary endothelial function in experimental porcine hypercholesterolemia in the absence of any lipid lowering effects [13], which translates into preservation of myocardial perfusion response and coronary microvascular integrity during episodes of increased cardiac demand [14]. In accordance with these results, pravastatin was shown to improve coronary endothelial function in cynomolgus monkeys, which were pretreated with an atherogenic diet for 2 years, independent of serum lipoprotein concentrations [15]. Since endothelial dysfunction is characterized by an imbalance between vasodilating and vasoconstricting substances, with an impairment of EDRFs and a predominance of EDCFs, statin-induced improvement in endothelial function is likely achieved by both enhancement of vasodilator and attenuation of vasoconstrictor activity in the vascular wall.

Table 1. Clinical Assessment of Vascular Endothelial Function

Endothelium-dependent vasodilatory response- Drug response: acetylcholine (ACh) Shear stress: reactive hyperemia
1 Coronary artery: coronary angiography Conduit vessel: change in vessel diameter after ACh provocation assessed by quantitative coronary angiography Resistance vessel: change in coronary flow assessed by Doppler coronary flow velocimetry
2 Peripheral artery: coronary angiography Conduit vessel: change in vessel diameter of brachial artery after reactive hyperemia assessed by high resolution ultrasonography -------flow-mediated vasodilation (FMD) Resistance vessel: change in blood flow of brachial artery after ACh or reactive hyperemia assessed by strain gouge plethysmography

Clinical Assessment of Vascular Endothelial Function and Statins' Effects

Vascular endothelial function can be clinically assessed using physiological methods in both coronary as well as peripheral arteries (Table 1). Since the first description of endothelial dysfunction in atherosclerotic epicardial coronary arteries in 1986 by Ludmer and colleagues [16], invasive assessment of coronary endothelial function by quantitative coronary angiography along with graded intracoronary infusions of endothelium-dependent vasodilator such as acetylcholine (ACh), may be considered the "golden standard" for endothelial function testing. ACh has both actions of vasodilatation by promoting endothelial NO release and vasoconstriction by direct action to vascular smooth muscles. ACh dilates normal blood vessels that have intact endothelium, but paradoxically constricts the vessels if their endothelium is damaged [16-20]. Thus, the observation of ACh-induced vasomotion is very sound methods for evaluating vascular endothelial function. Coronary artery vasospasm is considered to be an ultimate feature of endothelial dysfunction [21] as well as to be caused by hyperconstriction of vascular smooth muscle [22]. Nowadays, intracoronary injection of ACh is also widely applied to diagnose vasospastic angina. On the other hand, coronary Doppler flow measurements with intracoronary injection of ACh provides us an information regarding endothelial function of coronary resistance vessel level [23,24].

During the last decade, less-invasive or non-invasive techniques such as the forearm blood flow measurement (FBF) by strain gauge plethysmography using the venous occlusion technique or flow-mediated vasodilation (FMD) by high resolution ultrasonography has developed to assess endothelium-dependent vascular function of the forearm arteries, namely peripheral vascular endothelial function. The former can detect mainly endothelial function of resistance vessel levels, while the latter mainly detects that of conduit artery levels [25]. Using strain gauge plethysmography, endothelial function can be evaluated by an ACh-induced blood flow increase or post-ischemic reactive hyperemia [26]. In this case, however, ACh must be administrated by direct intra-arterial infusion, which is somewhat invasive. On the other hand, post-ischemic reactive hyperemia is also mediated by endothelial NO [27]. A 5 minutes occlusion of flow to the upper extremity produced by inflation of a blood pressure cuff, followed by release of the occlusion, results in an immediate 5-10-fold increase in blood flow, namely reactive hyperemia. However, the technique using venous occlusion plethysmography with reactive hyperemia is a little complex. In contract, since reactive hyperemia increases the vessel wall shear stress in the proximal artery with subsequent blood flow-dependent vasodilation, FMD of the brachial artery is nowadays the most frequently used as a non-invasive surrogate of endothelial function. The non-invasive nature of this technique allows repeated measurements over time to study the effectiveness of various interventions that may affect vascular health, although there are several limitations such as reproductibility or confounding factors in this method. Uehara et al. [28] observed brachial artery FMD by non-invasive assessment was compared with coronary endothelial function testing by assessing the conduit vessel vasomotor response to ACh as a part of coronary angiography in the same patients and demonstrated that both were correlated. This result suggests that non-invasively measured peripheral arterial endothelial function can be a simple surrogate for coronary endothelial function.

Using above-mentioned techniques, statins' effects for vascular endothelial function of the coronary as well as peripheral arteries were extensively assessed in various clinical setting. We have investigated endothelial function of the brachial artery in patients with hypercholesterolemia by the assessment of reactive hyperemic blood flow increase using the strain gauge plethysmography and evaluated the effects of statin treatments on the endothelial function [29]. As a result, administration of 20 mg/day fluvastatin but not 10 mg/day pravastatin for 16 weeks improved the endothelial function (Figure 1). Jarvisalo et al. [30] observed the statin's effects on peripheral endothelial function by a study comparing FMD in a group of 23 men with coronary artery disease without lipid-lowering medication with that in 22 age- and blood pressure-matched coronary artery disease patients with similar lipid level but ongoing statin therapy. In this study, FMD of the brachial artery was significantly higher in patients receiving statins than in those without any treatment. Moreover, multivariate analysis revealed the statin use as the only significant predictor of the FMD. The statins' effects was also evaluated on the endothelial function of coronary microvasculatures. Egashira et al. [31] demonstrated using coronary Doppler flow measurements with intracoronary ACh injection an evidence that exercise-induced myocardial ischemia even in patients without hemodynamically relevant coronary artery disease is associated with impaired endothelium-dependent vasodilation of coronary resistance vessels, namely endothelial dysfunction of coronary microvessels and that pravastatin administration improved this pathological status. Strong evidence of a role for endothelial dysfunction as an independent predictor of cardiovascular event stems several studies investigating the presence of endothelial dysfunction in coronary as well as systemic circulations and prognosis [32-37]. Thus, statin's effects on the reduction of coronary events may be in part derived from the improvement in endothelial function.

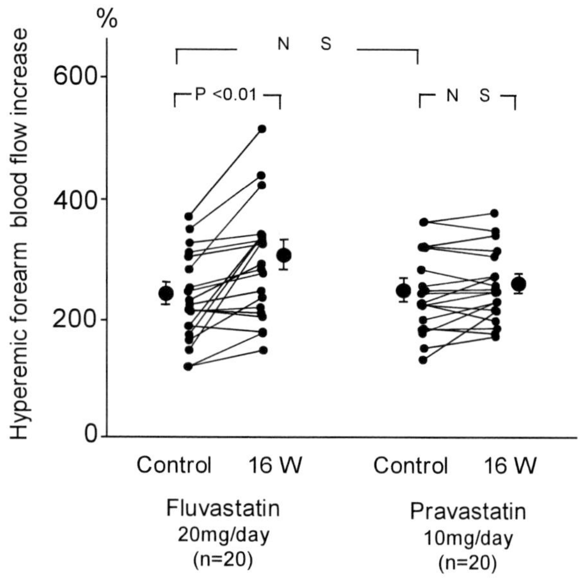

Figure 1. Effects of statins on vascular endothelial function. Fluvastatin but not pravastatin improved the endothelial function in patients with hypercholesterolemia by the assessment of reactive hyperemic blood flow increase using the strain gauge plethysmography.

Vascular Smooth Muscle Dysfunction and Statin

Vascular smooth muscle dysfunction seems to be classified two categories, impaired smooth muscle relaxation and smooth muscle cell proliferation. Endothelium-dependent vasodilation, i.e., endothelium-smooth muscle cell coupling also means vascular smooth muscle cell function in a broad sense. Thus, smooth muscle cell function cannot be separate from endothelial function and endothelial cell function assessment by above mentioned methods simultaneously assesses smooth muscle cell function, too. On the other hand, vasodilatory capacity after administration of NO donor such as sodium nitroprusside or nitroglycerin can assess the endothelium-independent vasodilation, impairment of which usually accompanied with smooth muscle cell proliferation [38]. In addition to improvement of endothelial function that seems to mean the improvement of smooth muscle cell function, statins also have direct action of endothelium-independent vasodilation by inhibiting inward L-type Ca++ current in smooth muscle cells [39]. It is required to check whether statin improves smooth muscle function in clinical setting. In the process of arteriosclerosis development, medial smooth muscle cell proliferates and migrate into the intima, accompanied with plaque formation. Thus, vascular smooth muscle cell proliferation, i.e., smooth muscle bulk, lead to neointimal thickening, meaning in some degree progression of evoked atherosclerosis. It has been known that statins inhibit proliferation and migration of smooth muscle cells by inducing apoptosis. Statins suppress c-fos gene expression by impairing insulin-like growth factor (IGF)/insulin signaling, also leading to inhibiting smooth muscle cell proliferation and migration.

Intimal-medial thickeness (IMT) of carotid artery observed by ultrasonogram is widely used to assess this stage of "Vascular Failure". Several reports demonstrated a relationship between the severity of carotid artery IMT and that of coronary artery disease evaluated by coronary angiography or myocardial perfusion scintigraphy [40-43]. There are several evidences showing that progression of carotid artery IMT could predict outcomes of cardiovascular events including coronary artery disease. In addition, intravascular ultrasound (IVUS) imaging directly provides us information regarding intimal thickening also in the coronary artery even where we cannot detect angiographic luminal narrowing. Statins' effects for regression of carotid IMT have been extensively reported in various clinical trials, and thus, the surrogacy of carotid IMT with respect to statin therapy has been evident [44,45]. Nishioka et al. [46] investigated the effects of statins on intimal thickness of the coronary artery using the IVUS imaging in patients undergoing coronary stenting, focusing on the plaques of non-treated segments and observed statins' effect on regression of coronary plaques without positive vessel remodeling.

Inflammation in the Process of AtheroscleIosis

Recently, atherosclerosis has widely recognized as an inflammatory disease. Recent advances in basic science have established a fundamental role for inflammation in mediating all stages of atherosclerosis from initiation through progression to the plaque formation and ultimately, plaque rupture and subsequent thrombotic complications in acute coronary syndrome (ACS) [47-49]. Initiation and progression of atherosclerosis are characterized by recruitment of monocytes and T-lymphocytes to artery wall, which is promoted by interaction between leukocytes and vascular endothelial cells. Increased leukocyte-endothelial adhesion is present in hypercholesterolemia. A triggering event for this process is accumulation of oxidized low-density lipoprotein (LDL), which stimulates the overlying endothelial cells to produce a number of pro-inflammatory molecules, including adhesion molecules such as intercellular adhesion molecule-1 (ICAM-1), vascular cell adhesion molecule-1 (VCAM-1), P-selectin, or E-selectin, chemotactic protein such as monocyte chemotactic protein-1 (MCP-1), and growth factors such as macrophage colony-stimulating factor (M-CSF), resulting in the recruitment of monocytes to the vessel wall. Oxidized LDL has other effects, such as inhibiting the production of NO, an important mediator of vasodilation and expression of E-selectin. Among endothelial cell adhesion molecules likely to be important in the recruitment of leukocytes are ICAM-1, VCAM-1, P-selectin, E-selectin, and Platelet/endothelial cell adhesion molecule (PECAM)-1. Important adhesion molecules on monocytes include beta 2-integrin lymphocyte functional antigen (LFA)-1 (CD11a/CD18) or Mac-1 (CD11b/CD18), beta 1-integrin vary late antigen (VLA)-4, and PECAM-1. Monocytes recruited into the vessel wall are differenciated into macrophages by M-CSF stimulation and express scavenger receptor that take up highly oxidized LDL, leading to form cell formation. The recent studies have shown that the interaction of CD40 and its ligand CD40L (CD154) makes an important contribution to the development of advanced lesions [50,51]. Vulnerable plaques have rich lipid, thin fibrous cap and abundant infiltration of inflammatory cells, monocytes/macrophages and T-lymphocytes. Activated macrophages produce matrix metalloproteinases (MMPs) that result in matrix degradation, leading to plaque rupture [52]. Once plaques rupture, in the process of thrombus formation, tissue factor activates coagulation cascade and subsequently platelets and leukocytes; neutrophils as well as monocytes are activated and interacted [53,54]. Thus, the process of plaque rupture is characterized as acute inflammatory reaction. On the other hand, restenosis after percutaneous coronary intervention (PCI) is also triggered by inflammation in the PCI-site injured vessel wall [55]. In the post-PCI inflammatory process, cross talk between platelet surface P-selectin and leukocyte Mac-1 has a crucial role [56,57].

Clinical studies have shown that this emerging biology of inflammation in the process of atherosclerosis from its onset to plaque vulnerability applies directly to human patients by the measurement of various inflammatory markers including cytokines such as interleukin (IL)-6 or tumor necrosis factor (TNF)-alpha, adhesion molecules such as ICAM-1, VCAM-1, P-selectin, or E-selectin, and acute phase reactant proteins exist in downstream of them such as C-reactive protein (CRP) or serum amyloid A (SAA). Among these inflammatory markers, the most useful marker for clinical use seems to be the CRP [58]. Recent development of highly sensitive measurement method of CRP (hs-CRP) enables us to assess low-grade inflammation and the CRP has been established as an independent predictor of cardiovascular events in healthy individuals. On the other hand, elevated circulating CRP accompanies ACS, reflecting a primary inflammatory instigator of vulnerable plaque [59]. In patients with ACS, elevated CRP is associated with an adverse in-hospital and short-term prognosis that includes death, recurrent episodes of ACS or myocardial infarction [60]. Recent research on CRP has focused on its localization and production in various inflammatory lesions, especially in atherosclerotic plaques, in addition to its production in the liver. We measured CRP values in patients with coronary artery disease in both blood samples from just distal and just proximal to the lesion, and observed that CRP was higher in the distal blood samples than in the proximal samples. The trans-lesional gradient of CRP (distal CRP minus proximal CRP) as well as the proximal CRP was higher in unstable angina patients than in stable angina patients. The trans-lesional CRP gradient correlated with the proximal CRP, namely systemic CRP. Thus, we observe in part CRP derived from atherosclerotic plaque in peripheral blood measurement [61]. CRP is also increased after PCI with the maximum increase at 48 hr after the procedure, and the CRP at 72 hr can predict restenosis [62].

Anti-Inflammatory Actions of Statins

Among the statins' pleiotropic effects, anti-inflammatory properties might represent one of the most important actions. It was demonstrated that fluvastatin attenuates the leukocyte-endothelial cell adhesion responses in hypercholesterolemic rat model independently of any lipid lowering effect [63]. Recently, these findings were extended by demonstrating a significant reduction of leukocyte-endothelial cell interactions with simvastatin in a normocholesterolemic rat model in vivo, which was at least partly mediated by attenuated up-regulation of P-selectin on endothelial cells [64]. Moreover, it has been shown that the Rho is essential for integrin-mediated leukocyte adhesion to endothelial cells [65]. Because geranylgeranylation is required for Rho activation it can be speculated that statins modulate leukocyte-endothelial cell interactions in part by inhibition of geranylgeranylation of this protein [66]. Besides these anti-inflammatory effects on endothelial cell adhesion molecules, statins seem to exert similar effects on leukocyte adhesion molecules. Fluvastatin inhibits adhesive interaction between monocytes and human umbilical vein endothelial cells (HUVECs) by lowering the expression of LFA-1 on monocyte and ICAM-1 on HUVECs [67]. Addition to lovastatin to isolated human monocytes led to a significant reduction of surface expression of CD11b, which in turn was associated with decreased CD11b-dependent adhesion of monocytes to HUVECs [68]. Co-incubation with mevalonate, but not with LDL, reversed the lovastatin's effect, suggesting a cruicial role for early cholesterol precursors of the mevalonate pathway for the inhibitory effect of statins on integrin expression and leukocyte-endothelial cell interactions. Furthermore, it has been demonstrated that treatment with simvastatin is associated with attenuation of CD18 up-regulation in neutrophils in response to stimulation with leukotriene B4 (LTB4) in normocholesterolemic rats [69]. Recently, cerivastatin was shown to reduce monocyte adhesion to endothelium under physiological flow condition via downregulation of integrin adhesion molecules, CD11a, CD18 and VLA-4, and inhibition of actin polymerization via prevention of Rho translocation to the membrane [70]. Thus, statins may affect leukocyte-endothelial cell interactions by various mechanisms, which depend on their ability to inhibit HMG-CoA reductase but are independent of cellular cholesterol biosynthesis. Lovastatin blocks LFA-1 mediated adhesion and costimmulation of lymphocytes via direct binding to a specific site within LFA-1 [71], suggesting a novel mechanism of action, which is unrelated to statin-mediated inhibition of HMG-CoA reductase as found to contribute to the anti-inflammatory potential of statins.

Another anti-inflammatory action demonstrated for several statins is the reduction of the production of pro-inflammatory cytokines. In one study [72] fluvastatin and pravastatin were found to significantly inhibit angiotensin II-induced secretion of interleukin (IL)-6 in cultured human smooth muscle cells, whereas in another study [73] fluvastatin and simvastatin but not pravastatin reduced production of IL-6 and IL-1 beta in HUVECs. Moreover, it was found that the reduction of the expression of several pro-inflammatory mediators, such as IL-6 and MCP-1, exerted by lovastatin in an in-vivo model of local acute inflammation is dependent on the impairment of the biosynthyesis of non-sterol derivatives arising from the mevalonate pathway [74]. The observation that similar dose of atorvastatin and pravastatin produced a similar reduction of MCP-1 expression in different arterial vascular beds despite significant differences in their plasma lipid lowering potential in hypercholesterolemic pigs further supports the existence of such lipid-independent anti-inflammatory effects of statins in-vivo [75].

Atorvastatin, lovastatin and pravastatin were shown to suppress T-cell responses, repressing interferon (IFN) gamma-induced expression of major histocompatibility complex class II (MHC-II) molecules on various cell types [76]. This effect, suggesting an immunomodulatory role for statins, was limited to antigen-presenting cells requiring costimmulation by IFN-gamma, whereas antigen-presenting cells constitutively expressing MHC-II, such as B-cells and dendric cells, were not affected. Further evidence for an immunomodulatoty role for statins stems from a study demonstrating that pravastatin may exert a synergistic effect with cyclosporine regarding the inhibition of cytotoxic T-cell activity in vitro [77].

Statins' anti-inflammatory qualities may also contribute to stability of the vulnerable plaques. Simvastastin was demonstrated to dose-dependently inhibit migration and MMP-9 secretion of the human monocytic cell line THP-1, an effect that was reversed by the simultaneous addition of mevalonate and its derivates, farnesylpyrophosphate (FPP) and geranylgeranylpyrophosphate (GGPP) [78]. Also, fluvastatin and simvastatin decrease secretion of MMP-9 by human and mouse macrophages in culture by their inhibitory action on the mevalonate pathway [79]. These statins' inhibiting actions on MMPs may lead to prevention of plaque rupture and suggest a potential therapeutic paradigm of acute coronary syndrome.

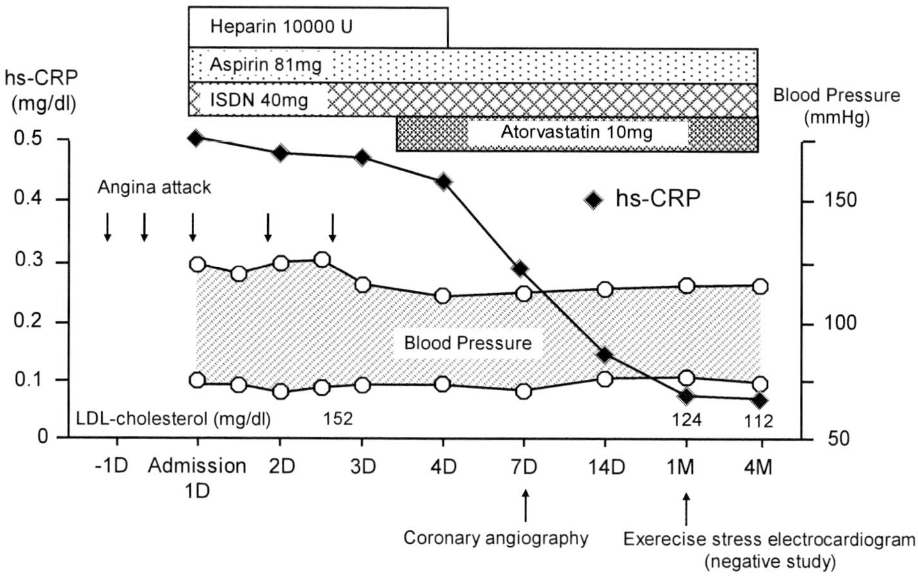

Figure 2. A case of 65 year-old male patient with unstable angina.

Statins' anti-inflammatory actions are also demonstrated a large clinical trial by the measurement of CRP. The MIRACL trial [7], as mentioned in the former chapter, evaluated the effects of aggressive lipid lowering by 80 mg/day of atorvastatin within 24-96 hrs from the onset in 3086 ACS patients. In this trial, CRP was measured as a surrogate marker and the result indicated that CRP was 34 % lower with atorvastatin than with placebo. Effects on CRP levels in human are investigated for various statins. Jialal et al. [80] tested the effects of 3 statins, simvastatin (20 mg/day), pravastatin (40 mg/day), and atorvastatin (10 mg/day), on levels of CRP in a randomized, double-blind, crossover trial of 22 patients with combined hyperlipidemia. The 3 statins similarly reduced CRP levels without no significant effects on either plasma interleukin-6 or interleukin-6 soluble receptor levels. There was no relationship between reductions in CRP and LDL cholesterol. Figure 2 shows a case of 65 year-old male patient with unstable angina. He admitted to hour hospital after 3 times new onset angina attack. After admission, his angina attack repeatedly appeared despite the intensive treatment with heparin, aspirin and nitrates. As serum LDL-cholesterol showed high level as 152 mg/dl, atorvastatin 10 mg/dl was prescribed, and then his angina disappeared. The LDL-cholesterol level decreased to 124 mg/dl. During his course, CRP level was monitored. The CRP level showed over 0.5 mg/dl at the begining but gradually decreased under 0.1 mg/dl, in association with stabilization of the symptom. On the 7th hospital day, coronary angiography showed a stenotic lesion in right coronary artery, and an intravascular ultrasound imaging showed a plaque with lipid core in this lesion (Figure 3). As a result, this patient avoided to undergo PCI because ischemic evidence was absent symptomatically or in exercise stress electrocardiography. In this patient, atorvastatin's pleiotropic effect, especially anti-inflammatory effect in addition to lipid lowering might result in plaque stabilization and favorable outcome.

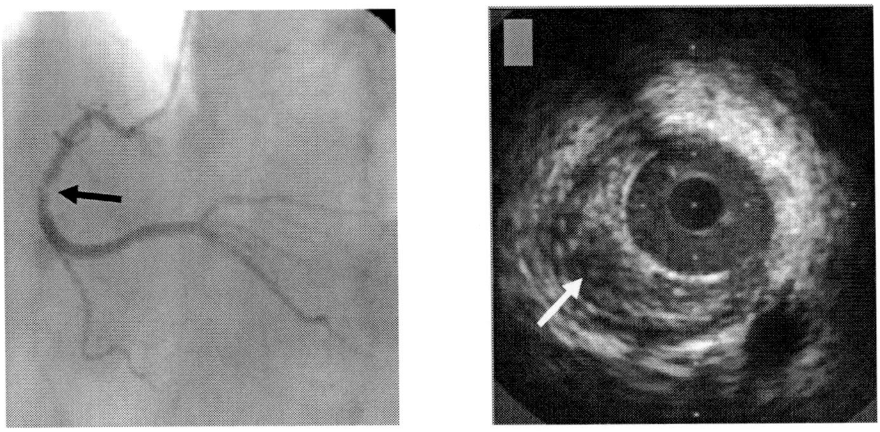

Figure 3. Coronary angiography showed a stenotic lesion in right coronary artery (left, arrow), and an intravascular ultrasound imaging showed a plaque with lipid core in this lesion (right, arrow).

Oxidative Stress in Athergenesis

Increasing numbers of studies have demonstrated that oxidative stress, i.e. dysregulation of cellular redox state, plays a pivotal role in the pathogenesis of atherosclerosis, especially vascular endothelial dysfunction [81]. Superoxide anion is formed by univalent reduction of molecular oxygen. Although several enzymes are involved in the generation of superoxide anion, including xanthine oxidase, NADH/NADPH oxidase, lipoxygenase and nitric oxide synthase, one of the largest factories producing superoxide anion in vivo is the mitochondrion [82]. Via spontaneous or enzymatically catalyzed dismutation, superoxide anion is reduced to hydrogen peroxide. Transition metal, such as iron or copper, -catalyzed interaction with hydrogen peroxide produces highly toxic hydroxyl radicals. Reactive oxygen specimen (ROS) have detrimental effect on vascular function through several mechanisms. First, as their direct effect, reactive oxygen species, especially hydroxyl radicals, injure cell membrane and nuclei. Second, by interacting with endogeneous vasoactive mediators formed in endothelial cells, reactive oxygen species modulate vasomotion and atherogenic process. Third, reactive oxygen species peroxide lipid components, leading to formation of oxidized LDL, which is one of the key mediators of atherosclerosis [83]. Whereas native LDL does not cause cholesterol ester accumulation in macrophages, modified LDL by oxidation does [84]. Oxidized-LDL has also been implicated in other mechanisms potentially involved in the development of atherosclerosis, i.e., cytotoxic or chemotactic actions on monocytes and inhibition of macrophage motility [85].

Table 2. Oxidative stress marker

Thiobarbituric acid reactive substance (TBARS)
8--iso-prostaglandin F2α
8-hydroxy-2'-deoxyguanosine (8-OHdG)
Oxidized LDL
Malonic dialdehyde -modified LDL (MDA-LDL)
Anti-oxidized LDL antibody
Oxidized α1-antitrypsin
Oxidized LDL receptor (LOX-1)
Asymmetric dimethylarginine (ADMA)
Tetrahydrobiopterin (BH4)
etc

There are a number of clinical markers suggested to assess oxidative stress status (Table 2). Thiobarbituric acid reactive substance (TBARS) is the lipid peroxide derived from unsaturated fatty acids. Isoprotanes such as 8-iso-prostaglandin (PG) F2 alpha or 8-epi-PGF2 alpha are compound formed in vivo by nonenzymatic free radical catalyzed peroxidation of arachidonic acid. TBARS and isoprostanes are both markers for non-specific lipid peroxidation. Elevated urinary 8-iso-PGF2 alpha levels are found in patients with elevated postprandial remnant lipoproteins or in smokers, even in passive smokers. 8-hydroxy-2'-deoxyguanosine (8-OHdG) is a degradation product of DNA oxidation and its measurement provides information on various degrees of oxidative stress in the DNA level. Serum 8-OHdG levels are elevated in smokers and the smoking abstension reduced the levels [86]. Among various lipids peroxidazed, oxidatively modified LDL directly involve in the athelosclerosis process. Thus, the direct measurement of oxidized LDL (Ox-LDL) provides important information regarding oxidative stress in atherosclerosis. On the other hand, specific immunological epitopes expressed on Ox-LDL were found in atherosclerotic lesions both in animals with experimental atherosclerosis and in humans. Expression of such epitopes in vitro can be generated by various procedures, including incubation with endothelial cells and macrophages, oxidation in the presence of copper ions and treatment with malondialdehyde [87]. Ox-LDL can interact with scavenger receptors of monocyte-derived macrophages. It is suggested that these interactions can induce formation of anti-Ox-LDL. Therefore, anti-Ox-LDL can be considered to be a marker of LDL oxidation at the level of tissues or cells. Since anti-Ox-LDL titer in serum of the patients with hypercholesterolemia is not correlated with LDL-cholesterol level but with 8-OHdG, the anti-Ox LDL also seems to be a marker of oxidative DNA damage in dyslipidemic patients [88]. The titer of anti-Ox LDL is associated with short-term lesion progression or regression of atherosclerotic coronary artery disease [89] as well as peripheral artery disease [90,91], or higher in patients with ACS than in patients with stable angina [92].

Anti-Oxydative Effects of Statins

Some kinds of statins have been considered to have anti-oxidative effects. Wilson et al. [93] demonstrated that simvastatin decreased plasma levels of 8-epi-PGF2 alpha and malondialdehyde, both markers of which indicate increased oxidative stress in vivo, in a model of experimental hypercholesterolemia, independently of lipid lowering effect. In addition, atorvastatin, pravastatin and cerivastatin may inhibit the NADPH oxidase-dependent superoxide anion formation in endothelial cells by preventing isoprenylation of the small GTP-binding protein Rac, which is essential for NADPH activation [94-96]. Atorvastatin increases catalase expression both in rat VSMCs in vitro and in normocholesterolemic spontaneously hypertensive rats in vivo [96]. A dose-dependent inhibitory effect on LDL oxidation was demonstrated for simvastatin [97]. Similarly, chronic administration of fluvastatin or lovastatin in hypercholesterolemic patients was shown to reduce the ex-vivo susceptibility of LDL to oxidation, which was thought to be partly mediated by direct binding of the drugs to the phospholipids fraction of LDL [98]. Among various statins, fluvastatin is thought to be the most powerful anti-oxidant. Differently from other statins, fluvastatin has lipid-independent strong radical scavenging action and reduces superoxide anion formation both in vitro and in vivo [99-101]. Fluvastatin has an indole ring in its structure, which is believed to be important for manifestation of this action. We demonstrated that fluvastatin 20 mg/day reduced anti-Ox-LDL titer and serum 8-OHdG levels in hypercholesterolemic patients (Figure 4) and the reduction of anti-Ox-LDL titer by fluvastatin was associated with that of serum 8-OHdG levels [102] In addition, we also demonstrated that anti-oxidative effect of fluvastatin attenuated nitrate tolerance, which is considered in part to occur in association with oxidative stress, in patients with coronary artery disease and dyslipidemia who had been receiving organic nitrates over long period [103]. It is noteworthy that because NO and superoxide anion interact chemically to neutralize each other, an increase in the local concentration of superoxide anion is associated with a decrease in the concentration of biologically active NO [94]. Thus, the antioxidant properties of statins may potentiate their effect on NO bioavailability.

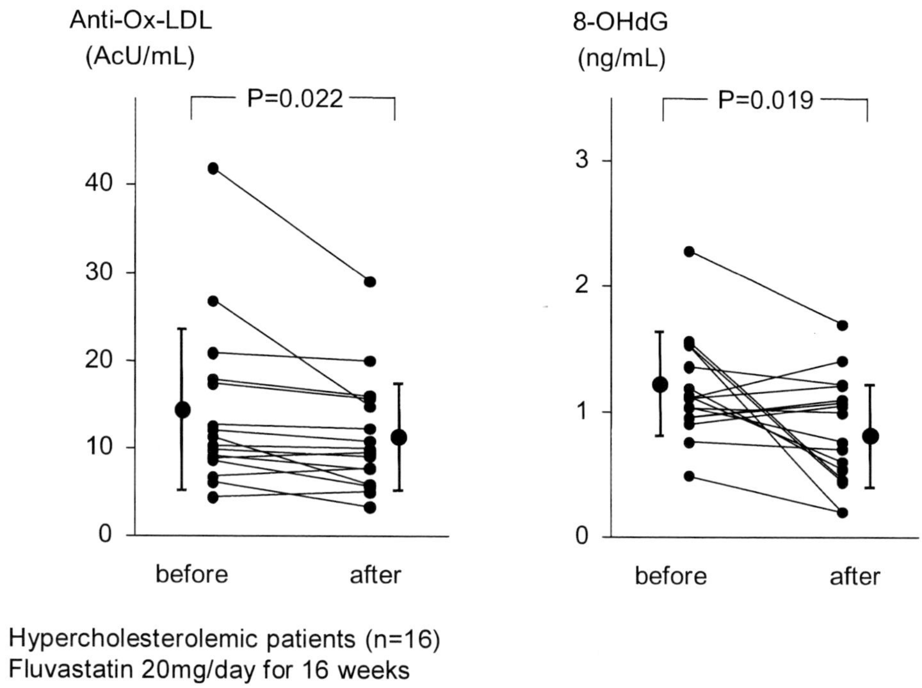

Figure 4. Anti-oxidative effect of fluvastatin. Fluvastatin decreased anti-Ox-LDL titer and 8-OHdG levels in hypercholesterolemic patients. anti-Ox-LDL=antibody against oxidized low-density lipoprotein 8-OHdG=8-hydroxy-2'-deoxyguanosine.

Effects of Statins on Platelets Activation and Thrombogenesis

Platelete activation and thrombogenesis in the process plaque of disruption is a key event in acute coronary syndrome. Since the endothelial cells of the coronary plaques is eroded or denuded, blood is directly exposed to procoagulant elements on subendothelial tissues, triggering the coagulation cascade, platelet aggregation, and fibrin deposition, which may lead to occlusive or semiocclusive thrombus formation. This thrombogenic process determines clinical outcome, which may vary from event free to myocardial infarction or even sudden death.

Statins inhibit platelet aggregation in part by lipid lowering because changes the cholesterol content of platelet membranes alters membrane fluidity [104]. Statins' effects on endothelial NO production may also inhibit platelet aggregation independently of lipid lowering. Atorvastatin has been shown to upregulate eNOs in platelets and to decrease platelet activation in vivo without lowering cholesterol levels [105]. Moreover, statins' effect on decreasing isoprostanes, such as 8-iso-PGF2 alpha or 8-epi-PGF2 alpha, which are potent platelet activators as well as oxidative stress markers, may also lead to inhibiting platelet aggregation [106]. In addition to antiplatelet actions, statins may also enhance antithrombic activities mediated by inhibiting coagulation system. For example, simvastatin, fluvastatin and cerivastatin were shown to reduce expression of tissue factor (TF) in cultured human monocytes/macrophages. This effect was reversed by co-incubation with mevalonate or all-trans-geranylgeraniol but not cholesterol, indicating its dependence on statin-induced reduction of intracellular GGPP biosynthesis independently of lipid lowering [107]. Therefore, statins may shift the fibrinolytic balance within the vessel wall towards increased fibrinolytic activity. Simvastatin also inhibits the expression of plasminogen activator inhibitor-1 (PAI-1) from human VSMCs and endothelial cells, while it increases the expression of tissue-type plasminogen activator (t-PA) from endothelial cells [108]. The upregulation of the fibrinolytic potential of endothelial cells is demonstrated in a study showing that lovastatin increases t-PA activity and decreases PAI-1 activity in a rat endothelial cell line in a time- and concentration-dependent manner. In this study, the lovastatin-induced modification of the endothelial fibrinolytic activity was found to be caused by inhibition of Rho geranylgeranylation and disruption of cellular actin filaments [109].

The anti-platelet, anti-coagulatory, and pro-fibrinolytic effects of statins observed in vitro suggests an important role for statins in the therapy of acute coronary syndrome associated with an increased thrombogenesis.

Effects of Statins on Neovascularization

Angiogenesis is a formation of new blood vessels by germination from the preexisting vessels, which is a favorable mechanism to restore the blood flow in ischemic diseases such as coronary artery disease and peripheral artery occlusive diseases. On the other hand, excessive vascularization is also considered to be associated with atherosclerotic plaque formation [110]. Angiogenesis shows very complex process that depends on the interaction of both pro- and anti-angiogenic molecules to form functional vessels [111]. Statins are considered to have both pro- and anti-angiogenic actions. Statins' anti-angiogenic actions are represented by inhibition of endothelial cell migration, possibly through inhibiting Rho geranylgeranylation [112]. Simvastatin prevents vasa vasorum neovascularization, indicating the inhibitory effects of statins on angiogenesis [113], but this seems to act to inhibit atherosclerotic plaque progression. Vascular endothelial growth factor (VEGF) is one of the key growth factors involved in angiogenesis [114]. Statins' actions on VEGF are controversial. Fluvastatin results in a significant reduction of VEGF levels that are elevated in patients with hyperlipidemia in both lipid-dependent and –independent fashions [115]. Since elevated level of VEGF is associated with enhanced atherosclerotic plaque progression and increased plaque macrophage, the VEGF reduction by fluvastatin possibly leads to the prevention of plaque progression. Conversely, statins are also reported to enhance VEGF secretion from cultured VSMCs, leading to pro-angiogenetic actions [116]. On the other hand, atorvastatin therapy is demonstrated to lead to an early increase in the number and the functional activity of circulating endothelial progenitor cells (EPCs) in patients with stable coronary artery disease [117]. EPCs are bone marrow derived cells that home to neovascularization sites and differentiate into endothelial cells in situ. Recently, statins have been shown to induce EPCs mobilization from the bone marrow and EPCs differentiation into endothelial cells via the serine/threonine kinase Akt signaling pathway [118,119]. Since enhancement of eNOS and NO production by activating the Akt in endothelial cells is associated with the induction of angiogenesis, statins may also promote angiogenesis through NO production.

Although statins have the potential to both inhibit and promote neovascularization, both inhibiting and promoting actions seem to protect ischemia. A different angiogenic response to statins may depend on the difference of vascular beds for their affected sites and different pathological state.

Effects of Statins on Myocardial Protection

Early and effective reperfusion is a key factor minimizing myocardial injury after an acute coronary event. However, reperfusion itself may promote an inflammatory response and enhance myocardial injury [120,121]. During reperfusion, activated leukocytes infiltrate the myocardium, releasing proteases, proinflammatory cytokines, and oxygen-derived free radicals, thereby increasing vascular endothelium and cardiomyocyte damage [122,123]. In an animal model, pretreatment of statins are shown to reduce reperfusion injury and ventricular dysfunction. In addition, statin-treated rats showed lower adherence of neutrophils to vascular endothelium and lower infiltration in the ischemic myocardium. This attenuation in neutrophil-endothelial interaction seems to be the consequence of a reduction in the expression of adhesion molecules from endothelial cells and an inhibition of neutrophil activation after statin treatment. The myocardial protective effects of statins is also detectable in the absence of neutrophils. The addition of the active form of simvastatin to the perfusion medium of isolated rat hearts reduces ischemia/reperfusion injury [124]. In addition, statin treatment partially prevents endothelial NO synthase reduction induced by ischemia/reperfusion injury; this cardioprotective effect of statins is completely abolished by simultaneous treatment with an NOS inhibitor. These findings in animal models suggest that acute treatment with statins could potentially attenuate ischemia/reperfusion injury by a lipid-independent mechanism. Despite the relevance of these findings, further studies are awaited to clarify the lipid-independent mechanism for cardioprotection.

Myocardial hypertrophy, fibrosis and left ventricular remodeling after myocardial infarction are key determinants of deterioration of long term prognosis. Statins inhibit intracellular signaling pathways involved in cardiac hypertrophy and fibrosis including downregulation of the activity of small GTP-binding proteins of the Rho family [125,126]. In addition, statins may modulate the remodeling process. Recently, it has been demonstrated that statin therapy improves cardiac function and symptoms in patients even with non-ischemic cardiomyopathy [127]. Myocardial MMPs are increased during the development of dilated cardiomyopathy. Since statins suppress growth of macrophages expressing MMPs, it is possible that statins suppress ventricular remodeling through inhibitory effects on MMPs in addition to the reduction of inflammatory cytokines [128].

Future Aspects of Statin Therapy for Coronary Artery Disease

Nowadays we can see from a large number of clinical trials that statins are established drugs for preventing coronary events. Considering their favorable direct effects on blood vessels independently of lipid-lowering effects, statins should be aggressively applied for treatment of coronary artery disease even in patients without dyslipidemia. Among multiple actions, the most unique one may be stabilization of vulnerable plaques. Since statins have recently been recognized to have acute pharmacological actions, we should start to prescribe them as early as possible in patients with acute coronary syndrome, prior to PCI, if possible, to improve long-term prognosis after PCI. Statins will be also used in patients with effort angina to target inhibition of coronary artery lesion progression or to expect the improvement of nitrate tolerance. Furthermore, statins' effects of improvement of vascular endothelial function or vascular smooth relaxation make us to expect their application for vasospastic angina. Statins' inhibition of SMCs and anti-inflammatory actions may prevent post-PCI restenosis, and statin-coating stents are now under investigation. However, the most hopeful endpoint may be primary prevention of coronary artery disease onset by the use of statins in patients independently of their cholesterol levels. To apply statin therapies for these settings, we must do several efforts. First, we must establish and standardize diagnostic methods for clinically detecting 'Vascular Failure' to assess the effectiveness of statin therapies. The degree of 'Vascular Failure' will be a surrogate endpoint to evaluate the vascular pleiotropic effects of statins. Next, we must re-evaluate pleiotropic effects of each statin to know how to use properly. In addition, development of new statins that has less unfavorable adverse effects such as rhabdomyolysis and renal dysfunction would be needed so that we use them more readily. In future, statin therapy would be more extensively applied even in normolipidemic patients if they have other conventional risk factors such as hypertension, diabetes mellitus, or others. Furthermore, we may need to use statins to intervene in earlier stage risk factors such as postprandial hyperlipidemia or hyperglycemia, insulin resistant state, masked hypertension, or metabolic syndrome to further reduce mortality or morbidity of coronary artery disease.

References

[1] West of Scotland Coronary Prevention Study Group. Influence of pravastatin and plasma lipid levels on clinical events in the West of Scotland Coronary Prevention Study (WOSCOPS). *Circulation* 1998; 97: 1440-5.

[2] Downs JR, Clearfield M, Weis S, et al. Primary prevention of acute coronary events with lovastatin in men and women with average cholesterol levels: results of AFCAPS/TexCAPS. Air Force/Texas Coronary Atherosclerosis Prevention Study. *JAMA* 1998; 279: 1615-22.

[3] Scandinavian Simvastatin Survival Study Group. Randomized trial of cholesterol lowering 4444 patients with coronary heart disease: the Scandinavian Simvastatin Survival Study (4S). *Lancet* 1994; 344: 1383-9.

[4] Marschner IC, Colquhoun D, Simes RJ, et al. Long-term risk stratification for survivors of acute coronary syndromes. Results from the Long-term Intervention with Pravastatin in Ischemic Disease (LIPID) Study. LIPID Study Investigators. *J Am Coll Cardiol.* 2001; 38: 56-63.

[5] Sacks FM, Pfeffer MA, Moye' L, et al. Rationale and design of a secondary prevention trial of lowering normal plasma cholesterol levels after acute myocardial infarction: the Cholesterol and Recurrent Events trial (CARE). *Am J Cardiol* 1991; 68: 1436-46.

[6] Mikhaildis DP, Wierzbiki AS, Reynolds TM. Is a mechanical or metabolic approarch superior in the treatment of coronary disease? Results of the atorvastatin versus revascularization (AVERT) trial. *Eur Heart J* 2001; 22: 972-3.

[7] Kinlay S, Schwartz GG, Olsson AG, et al. High-dose atorvastatin enhances the decline in inflammatory markers in patients with acute coronary syndromes in the MIRACL study. *Circulation.* 2003; 108: 1560-9.

[8] Serruys PW, de Feyter P, Macaya C, et al. Lescol Intervention Prevention Study (LIPS) Investigators. Fluvastatin for prevention of cardiac events following successful first percutaneous coronary intervention: a randomized controlled trial. *JAMA.* 2002; 287: 3215-22.

[9] Luscher TF, Barton M. Biology of the endothelium. *Clin Cardiol* 1997; 20: 3-10.

[10] Lerman A, Burnet JC Jr. Intact and altered endothelium in regulation of vasomotion. *Circulation* 1992; 86(suppl III): III-12-II-19.

[11] Anderson TJ. Assessment and treatment of endothelial dysfunction in humans. *J Am Coll Cardiol* 1999; 34: 631-638.

[12] Bourcier T, Libby P: HMG CoA reductase inhibitors reduce plasminogen activator inhibitor-1 expression by human vascular smooth muscle and endothelial cells. *Arterioscler Thromb Vasc Biol* 2000; 20: 556-62.

[13] Wilson SH, Simari RD, Best PJM et al. Simvastatin preserves coronary endothelial function in hypercholesterolemia in the absence of lipid lowering. *Arterioscler Thromb Vasc Biol* 2001; 21: 172-8.

[14] Bonetti PO, Wilson SH, Rodriguez-Porcel M et al. Simvastatin preserves myocardial perfusion and coronary microvascular permeability in experimental hypercholesterolemia independent of lipid lowering. *J Am Coll Cardiol* 2002; 40: 546-54.

[15] Williams JK, Sukhova GK, Herrington DM, et al, Pravastatin has cholesterol-lowering independent effects on the artery wall of atherosclerotic monkeys. *J Am Coll Cardiol* 1998; 31: 684-91.

[16] Ludmer PL, Selwin AP, Shook TL, et al. Paradoxical vasoconstriction induced by acetylcholine in atherosclerotic coronary arteries. *N Engl J Med* 1986; 315: 1046-51.

[17] Furchgott RF. Role of endothelium in responses of vascular smooth muscle. *Circ Res* 1983; 53: 984-9.

[18] Vanhoutte PM, Rimele TJ. Role of endothelium in the control of vascular endothelial function. *J Physiol (Paris)* 1983; 78: 681-9.

[19] Werns SW, Walton JA, Hsia HH, et al. Evidence of endothelial dysfunction in angigraphically normal coronary arteries of patients with coronary artery disease. *Circulation* 1989; 79: 287-91.

[20] Horio Y, Yasue H, Rokutanda M, et al. Effects of intracoronary injection of acetylcholine on coronary artery diameter. *Am J Cardiol* 1986; 57: 984-9.

[21] Yasue H, Omote S, Takizawa A, et al. Coronary arterial spasm in ischemic heart disease and its pathogenesis: a review. *Circ Res* 1983; 52(supple I): I-147-52.

[22] Shimokawa H. Rho-kinase as a novel therapeutic target in treatment of cardiovascular diseases. *J Cardivasc Pharmacol* 2002; 39: 319-27.

[23] Egashira K, Inou T, Hirooka K, et al. Evidence of impaired endothelium-dependent coronary vasodilatation in patients with angina pectoris and normal coronary angiogram.

[24] Inoue T, Sakai Y, Morooka S, et al. Vasodilatory capacity of coronary resistance vessels in dilated cardiomyopathy. *Am Heart J* 1994; 127: 376-81.

[25] Wu HD, Katz SD, Beniaminovitz A, et al. Assessment of endothelium-mediatedvasodilation of the peripheral circulation by transcutaneous ultrasonography and venous occlusion plethysmography. *Heart Vessels* 1999; 14: 143-8

[26] Higashi Y, Yoshizumi M. New method to evaluate endothelial function: method for assessing endothelial function in humans using a strain-gauge plethysmography: nitric oxide-dependent and-independent vasodilation. *J Pharmacol Sci* 2003; 93: 399-404

[27] Olsen SP, Clapham DE, Davis PF: Hemodynamic shear stress activates a K+ current in endothelial cells. *Nature* 1998; 331: 168-70

[28] Uehara, A, Gerhard MD, Meredith IT, et al. Close relationship of endothelial dysfunction in coronary and brachial artery. *Circulation* 1993; 88: I-618

[29] Inoue T, Hayashi M, Takayanagi K, et al. Lipid-lowering therapy with fluvastatin inhibits oxidative modification of low-density lipoprotein and improves vascular endothelial function in hypercholesterolemic patients. *Atherosclerosis* 2002; 160: 369-76.

[30] Jarvisalo MJ, Toikka JO, Vasankari T, et al. HMG CoA reductase inhibitors are related to improved systemic endothelial function in coronary artery disease

[31] Egashira K, Inou T, Hirooka K, et al. Evidence of impaired endothelium-dependent coronary vasodilatation in patients with angina pectoris and normal coronary angiogram.

[32] Al Suwaidi J, Hamasaki S, Higano ST, et al. Long-term follow-up of patients with mild coronary artery disease and endothelial dysfunction. *Circulation* 2000; 101: 948-54.

[33] Schächinger V, Britten MB, Zeiher AM. Prognostic impact of coronary vasodilator dysfunction on adverse long-term outcome of coronary heart disease. *Circulation* 2000; 101: 1899-906.

[34] Halcox JPJ, Schenche WH, Zalos G, et al. Prognostic value of coronary vascular endothelial dysfunction. *Circulation* 2002; 106: 653-8.

[35] Particone F, Ceravolo R, Pujia A, et al. Prognostic significance of endeothelial dysfunction in hypertensive patients. *Circulation* 2001; 104: 191-6.

[36] Heitzer T, Schlinzig T, Krohn K, et al. Endothelial dysfunction, oxidative stress, and risk of cardiovascular events in patients with coronary artery disease. *Circulation* 2001; 104: 2673-8.

[37] Gokce N, Kerney JF Jr, Hunter LM, et al. Risk stratification for postoperative cardiovascular events via noninvasive assessment of endothelial function: a prospective study. *Circulation* 2002; 105: 1567-72.

[38] Higashi Y, Sasaki S, Nakagawa K, et al. Excess norepinephrine impairs both endothelium-dependent and –independent vasodilation in patients with pheochromocytoma: a comparison before and after adrenalectomy. *Hypertension* 2002; 39: 513-8.

[39] Bergdahl A, Persson E, Hellstrand P, Sward K. Lovastatin induces relaxation and inhibits L-type Ca(2+) current in the rat basilar artery. *Pharmacol Toxicol.* 2003; 93:128-34.

[40] Holaj R, Spacil J, Petrasek J, et al. Intima-media thickness of the common carotid artey is the significant predictor of angiographically proven coronary artery disease. *Can J Cardiol* 2003; 19: 670-6.

[41] Matsushima Y, Kawano H, Koide Y, et al. Relationship of carotid intims-media thickness, pulse wave velocity, and ankle brachial index to the severity of coronary artery atherosclerosis. *Clin Cardiol* 2004; 27: 629-34.

[42] Kablak-Ziembicka A, Tracz W, Przewlocki T, et al. Association of increased carotid intima-media thickness with the extent of coronary artery disease. *Heart* 2004; 90: 1286-90.

[43] Hallerstam S, Larsson PT, Zuber E, et al. Carotid atherosclerosis is correlated with extent and severity of coronary artery disease evaluated by myocardial perfusion scintigaphy. *Angiology* 2004; 55: 281-8.

[44] Taylor AJ, Kent SM, Flaherty PJ, et al. ARBITER: Arterial Biology for the Investigation of the Treatment Effects of Reducing Cholesterol: a randomized trial comparing the effects of atorvastatin and pravastatin on carotid intima medial thickness. *Circulation.* 2002;106: 2055-60.

[45] Smilde TJ, van Wissen S, Wollersheim H, et al. Effect of aggressive versus conventional lipid lowering on atherosclerosis progression in familial hypercholesterolaemia (ASAP): a prospective, randomised, double-blind trial. *Lancet.* 2001; 9256: 577-81.

[46] Nishioka H, Shimada K, Kataoka T, Hirose M, Asawa K, Hasegawa T, Yamashita H, Ehara S, Kamimori K, Sakamoto T, Kobayashi Y, Yoshimura T, Yoshiyama M, Takeuchi K, Yoshikawa J. Impact of HMG-CoA reductase inhibitors for non-treated coronary segments. *Osaka City Med J* 2004; 50: 61-8

[47] Libby P, Ridker PM, Maseri A. Inflammation and atherosclerosis. *Circulation* 2002; 105: 1135-43.

[48] Lusis AJ. Atherosclerosis. *Nature* 2004; 407: 233-41.

[49] Shishehbor MH, Bhatt DL. Inflammation and atherosclerosis. *Curr Atheroscler Reports* 2004; 6: 131-9.

[50] Schönbeck U, Sukhova GK, Shimizu K, et al. Inhibition of CD40 signaling limits evolution of established atherosclerosis in mice. *Proc Natl Acad Sci USA* 2000; 97: 7458-63.

[51] Henn V, Slupsky JR, Grafe M et al. CD40 ligand on activated platelets triggers an inflammatory reaction of endothelial cells.

[52] Virmani R, Burke AP, Farb A, et al. Pathology of the unstable plaque. *Prog Cardiovasc Dis* 2002; 44: 349-356.

[53] Ott I, Newmann FJ, Gawaz M et al. Increased neutrophil-platelet adhesion in patients with unstable angina. *Circulation* 1996; 94: 1239-1246.

[54] Furman MI, Benoit SE, Barnard MR, et al. Increased platelet reactivity and circulating monocyte-platelet aggregates in patients with stable coronary artery disease. *J Am Coll Cardiol* 1998; 31: 352-358.

[55] Toutouzas K, Colombo A, Srefanadis C. Inflammation and restenosis after percutaneous coronary interventions. *Eur Heart J* 2004; 25: 1679-1687.

[56] Welt FGP, Rogers C. Inflammation and restenosis in the stent era. Arterioscler Thromb Vasc Biol 2002; 22: 1769-1776.

[57] Inoue T, Uchida T, Yaguchi I, et al. Stent-induced expression and activation of the leukocyte integrin Mac-1 is associated with neointimal thickening and restenosis. *Circulation.* 2003; 107: 1757-1763.

[58] Ridker PM, Stampfer MJ, Rifai N. Novel risk factors for systemic atherosclerosis: a comparison of C-reactive protein, fibrinogen, homocysteine, lipoprotein (a), and standard cholesterol screening as predictors of peripheral arterial disease. *JAMA* 2001; 285: 2481-2485.

[59] Berk BC, Weintraub WS, Alexander RW. Elevation of C-reactive protein in "active" coronary artery disease. *Am J Cardiol.* 1990; 65: 168-172.

[60] Liuzzo G, Biasucci LM, Gallimore JR, et al. The prognostic value of C-reactive protein and serum amyloid a protein in severe unstable angina. *N Engl J Med.* 1994; 331: 417-424.

[61] Saito M, Inoue T, Hasegawa K, et al. C-reactive protein is derived from inflammatory vulnerable plaque in acute coronary syndrome. *Circulation* 2004; 110 (supple): III-697.

[62] Versaci F, Gaspardone A, Tomai F et al. Immunosuppressive therapy for the prevention of restenosis after coronary artery stent implantation. *J Am Coll Cardiol* 2002; 40: 1935-1942.

[63] Kimura M, Kurose I, Russell J, et al. Effects of fluvastatin on leukocyte-endothelial cell adhesion in hypercholesterolemic rats. *Arterioscler Thromb Vasc Biol* 1997; 17: 1521-6.

[64] Pruefer D, Scalia R, Lefer AM. Sinvastatin inhibits leukocyte-endothelial cell interactions and protects against inflammatory processes in normocholesterolemic rats. *Arterioscler Thromb Vasc Biol* 1999; 19: 2894-900.

[65] Wojciak-Stothard B, Williams L, Ridly AJ. Monocyte adhesion and spreading on human endothelial cells is dependent on Rho-regulated receptor clustering. *J Cell Biol* 1999; 145: 1293-307.

[66] Kwak BR, Mach F. Statins inhibit leukocyte recruitment –new evidence for their anti-inflammatory properties. *Aeterioscler Thromb Vasc Biol* 2001; 21: 1256-9

[67] Niwa S, Totsuka T, Hayashi S. Inhibitory effect of fluvastatin, an HMG-CoA reductase inhibitor, on the expression of adhesion molecules on human monocyte cell line. *Int J Immunopharmacol* 1996; 18: 669-75.

[68] Weber C, Erl W, Weber KSC, et al. HMG-CoA reductase inhibitors decrease CD11b expression and CD11b-dependent adhesion of monocytes to endothelium and reduce increased adhesiveness of monocytes isilated from patients with hypercholesterolemia. *J Am Coll Cardiol* 1997; 30: 1212-7.

[69] Lefer AM, Campbell B, Shin YK, et al. Simvastatin preserves the ischemic-reperfused myocardium in normocholesterolemic rat hearts. *Circulation* 1999; 100: 178-84.

[70] Yoshida M, Sawada T, Ishii H. HMG-CoA reductase inhibitor modulates monocyte-endothelial cell interaction under physiological flow conditions in vitro: involvement of Rho GTPase-dependent mechanism. *Arterioscler Thromb Vasc Biol* 2001; 21: 1165-71.

[71] Weitz-Schmidt G, Welzenbach K, Brinkmann V, et al. Statins selectively inhibit leukocyte function antigen-1 by binding to a novel regulatory integrin site. *Nat Med* 2001; 7: 687-92.

[72] Ikeda U, Ito T, Shimada K, et al. Statins and C-reactive protein. *Lancet* 1999; 353: 1274-5.

[73] Inoue I, Goto S, Mizotani K, et al. Lipophilic HMG-CoA reductase inhibitor as an anti-inflammatory effect: reduction of mRNA levels for interleukin-1β, interleukin-6, cyclooxygenase-2, and p22phox by regulation of peroxisome proliferators-activated receptor α PPARα) in primary endothelial cells. *Life Sci* 2000; 67: 863-76.

[74] Diamonde L, Albani D, Sottocorno M, et al. In vivo anti-inflammatory effect of statins is mediated by nonsterol mevalonate products. *Arterioscler Thromb Vasc Biol* 2001; 21: 1327-32.

[75] Martinez-Gonzalez J, Alfon J, Berrozpe M, et al. HMG-CoA reductase inhibitors reduce vascular monocyte chemotactic protein-1 expression in early lesions from hypercholesterolemic swine independently of their effect on plasma cholesterol levels. *Atherosclerosis* 2001; 159: 27-33.

[76] Kwak B, Mulhaupt F, Myit S, et al. Statins as newly recognized type of immunomodulator. *Nat Med* 2000; 6: 1399-402.

[77] Katzneison S, Wang XM, Chia D et al. The inhibitory effects of pravastatin on natural killer cell activity in vivo and on cytotoxic T lymphocyte activity in vitro. *J Heart Lung Transplant* 1998; 17: 335-40.

[78] Wong B, Lumma WC, Smith AM et al. Statins suppress THP-1 cell migration and secretion of matrix metalloproteinase 9 by inhibiting geranylgeranylation. *J Leukoc Biol* 2001; 69: 959-62.

[79] Bellosta S, Via D, Canavesi P, et al. HMG-CoA reductase inhibitors reduce MMP-9 secretion my macrophages. *Arterioscler Thromb Vasc. Biol* 1998; 18: 1671-8.

[80] Jialal I, Stein D, Balis D, Grundy SM, Adams-Huet B, Devaraj S. Effect of hydroxymethyl glutaryl coenzyme a reductase inhibitor therapy on high sensitive C-reactive protein levels. *Circulatiion* 2001; 103: 1933-5

[81] Matsuoka H. Endothelial dysfunction associated with oxidative stress in human. *Diabetes Res Clin Pr* 2001; 54 (supple 2): S65-S72.

[82] Harrison DG. Cellular and molecular mechanism of endothelial cell dysfunction. *J Clin Invest* 1997; 100: 2153-2157.

[83] Steinberg D. Low density lipoprotein oxidation and its pathological significance. *J Biol Chem* 1997; 272: 20963-20962.

[84] Fogelman AM, Schechter I, Seager J, et al. Malondialdehyde alteration of low density lipoproteins leads to cholesterol ester accumulation in human monocyte-macrophages. *Proc Natl Acad Sci USA* 1980; 77: 2214-2218.

[85] Quinn MT, Parthasarathy S, Fong LG, et al. Oxidatively modified low density lipoproteins: a potential role in recruitment and retention of monocyte/macrophages during atherosclerosis. *Proc Natl Acad Sci USA* 1987; 84: 2995-2998.

[86] Inoue T, Hayashi M, Takayanagi K, et al. Oxidative DNA damage is induced by chronic cigarette smoking, but repaired by abstension. *J Health Sci* 2003; 29: 217-220.

[87] Palinski W, Ylä-Herttuala S, Rosenfeld ME et al. Antisera and monoclonal antibodies specific for epitopes generated during oxidative modification of low density lipoprotein. *Arteriosclerosis.* 1990 ; 10 : 325-35.

[88] Inoue T, Inoue K, Maeda H, et al. Immunological response to oxidized LDL occurs in association with oxidative DNA damage independently of serum LDL concentrations in dyslipidemic patients. *Clin Chim Acta* 2001; 305: 115-121.

[89] Inoue T, Uchida T, Kamishirado H, et al. Antibody against oxidized low density lipoprotein may predict progression or regression of atherosclerotic coronary artery disease. *J Am Coll Cardiol* 2001; 37: 1871-1876.

[90] Salonen JT, Ylä-Herttuala S, Yamamoto R et al. Autoantibody against oxidized LDL and progression of carotid atherosclerosis. *Lancet* 1992; 339: 883-887.

[91] Bergmark C, Wu R, de Faire RW, et al. Patients with early-onset peripheral vascular disease have increased levels of autoantibodies against oxidzed LDL. *Arterioscler Thromb Vasc Biol.* 1995; 15: 441-445.

[92] Inoue T, Uchida T, Kamishirado H, et al. Clinical significance of antibody against oxidized low density lipoprotein in patients with atherosclerotic coronary artery disease. *J Am Coll Cardiol* 2001; 37: 775-779.

[93] Wilson SH, Simari RD, Best PJM, et al. Sinvastatin preserves coronary endothelial function in hypercholesterolemia in the absence of lipid lowering. *Arterioscler Thromb Vasc Biol* 2001; 21: 122-8.

[94] Wagner AH, Köhler T, Rückschloss U, et al. Improvement of nitric oxide-dependent vasodilation by HMG-CoA reductase inhibitors through attenuation of endothelial superoxide anion formation. *Arterioscler Thromb Vasc Biol* 2000; 20: 61-9.

[95] Wassmann S, Laufs U, Bäumer AT, et al. HMG-CoA reductase inhibitors improve endothelial dysfunction in normocholesterolemic hypertension via reduced production of reactive oxygen species. *Hypertension* 2001; 37: 1450-7.

[96] Wassmann S, Laufs U, Müller K, et al. Cellular antioxidant effects of atorvastatin in vitro and in vivo. *Arterioscler Thromb Vasc Biol* 2002; 22: 300-5.

[97] Girona J, La Ville AE, Sota R, et al. Simvastatin decreases aldehyde production derived from lipoprotein oxidation. *Am J Cardiol* 1999; 83: 846-51.

[98] Aviram M, Hussein O, Rosenblat M, et al. Interaction of platelets, macrophages, and lipoproteins in hypercholesterolemia: antiatherogenic effects of HMG-CoA reductase inhibitor therapy. *J Cardiovasc Pharmacol* 1998; 31: 39-45.
[99] Suzumura K, Yasuhara M, Tanaka K, et al. An in vitro study of the hydroxyl radical scavenging property of fluvastatin, an HMG-CoA redactase inhibitor. *Chem Pharm Bull* 1999; 47: 1010-2.
[100] Yamamoto A, Hoshi K, Ichihara K. Fluvastatin, an inhibitor of 3-hydroxy-3-methylglutaryl-CoA reductase, scavenges free radicals and inhibits lipid peroxidation in rat liver microsome. *Eur J Pharmacol* 1998; 361: 143-9.
[101] Rikitake Y, Kawashima S, Takeshita S, et al. Anti-oxidative properties of fluvastatin, an HMG-CoA reductase inhibitor, contribute to prevention of atherosclerosis in cholesterol-fed rabbits. *Atherosclerosis* 2001; 154: 87-96.
[102] Inoue T, Inoue K, Maeda H, et al: Immunological response to oxidized LDL occurs in association with oxidative DNA damage independently of serum LDL levels in dyslipidemic patients. *Clin Chim Acta* 2001; 305: 115-21.
[103] Inoue T, Takayanagi K, Hayashi T, et al: Fluvastatin attenuate nitrate tolerance in patients with ischemic heart disease. *Int J Cardiol* 2003; 90: 181-8.
[104] Rosenson RS, Tangney CC. Antiatherothrombotic properties of statin-implications for cardiovascular event reduction. *JAMA* 1998; 279: 1643-50.
[105] Laufs U, Gertz K, Huang P, et al. Atorvastatin upregulates type III nitric oxide synthase in thrombocytes, decreases platelet activation, and protects from cerebral ischemia in normocholesterolemic mice. *Stroke* 2000; 31: 2437-49.
[106] Kearney D, Fitzgerald D. The anti-thrombotic effects of statins. *J Am Coll Cardiol* 1999; 33: 1305-7.
[107] Colli S, Eligini S, Lalli M, et al. Vastatins inhibit tissue factor in cultured human macrophages: a novel mechanism of protection against atherothrombosis. *Arterioscler Thromb Vasc Biol* 1997; 17: 265-72.
[108] Bourcier T, Libby P. HMG-CoA reductase inhibitors reduce plasminogen activator inhibitor-1 expression by human vascular smooth muscle and endothelial cells. *Arterioscler Thromb Vasc Biol* 2000; 20: 556-62.
[109] Essig M, Nguyen G, Prie D, et al. 3-hydroxy-3-methylglutaryl-CoA reductase inhibitors increase fibrinolytic activity in rat aortic endothelial cells. Arole of geranylgeranylation and Rho properties. *Circ Res* 1998; 83: 683-90.
[110] Carmeliet P, Jain RK. Angiogenesis in cancer and other disease. *Nature* 2000; 407: 249-57.
[111] Yancopoulous GD, Davis S, Gale NW, et al. Vascular-specific growth factors and blood vessel formation. *Nature* 2000; 407: 242-8.
[112] Vincent L, ChenW, Hong L, et al. Inhibition of endothelial cell migration by cerivastatin, an HMG-CoA reductase inhibition: contribution to its anti-angiogenic effect. *FEBS Lett* 2001; 495: 159-55.
[113] Wilson SH, Hermann J, Lerman LO, et al. Simvastatin preserves the structure of coronary adventitial vasa vasorum in experimental hypercholesterolemia independent of lipid lowering.

[114] Semenza GL. Regulation of hypoxia-induced angiogenesis: a chaperone escorts VEGF to the dance. *J Clin Invest* 2001; 108: 39-40.
[115] Blann AD, Belgore FM, Constans J, et al. Plasma vascular endothelial growth factor and its receptor Flt-1 in patients with hyperlipidemia and atherosclerosis and the effects of fluvastatin or ferofibrate. *Am J Cardiol* 2001; 87: 1160-3.
[116] Nishioka H, Shimada K, Kataoka T, et al. Impact of HMG-CoA reductase inhibitors for non-treated coronary segments. *Osaka City Med J* 2004; 50: 61-8.
[117] Vasa M, Fichischerer S, Adler K, et al. Increase in circulating endothelial progenitor cells by statin therapy in patients with stable coronary artery disease. *Circulation* 2001; 103: 2885-90.
[118] Dimmeler S, Aicher A, Vasa M, et al. HMG-CoA reductase inhibitors (statins) increase endothelial progenitor cells via the PI-3 kinase/Akt pathway. *J Clin Invest* 2001; 108: 391-7.
[119] Lievadot J, Satoshi M, Kureishi Y, et al. HMG-CoA reductase inhibitor mobilizes bone marrow-derived endothelial progenitor cells. *J Clin Invest* 2001; 108: 399-405.
[120] Forman MB, Puett DW, Virmani R, et al. Endothelial and myocardial injury during ischemia and reperfusion: pathogenesis and therapeutic implication. *J Am Coll Cardiol* 1989; 13: 450-459.
[121] Tsao PS, Aoki N, Lefer DJ, et al. Time course of endothelial dysfunction and myocardial injury during myocardial ischemia and reperfusion in the cat. *Circulation* 1990; 82: 1402-12.
[122] Weiss SJ. Tissue destruction by neutrophils. *N Engl J Med* 1989; 320: 363-376.
[123] Engler RL, Schmid-Schonbein GW, Pavelec RS. Leukocyte capillary plugging in myocardial ischemia and reperfusion in the dog. *Am J Pathol* 1983; 111: 98-111.
[124] Denapoli P, Antonio Taccardi A, Grilli A, et al. Simvastatin reduces reperfusion injury by modulating nitric oxide synthase expression: an ex vivo study in isolated working rat hearts. *Cardiovasc Res* 2001; 51: 283-93.
[125] Laufs U, Kilter M, Kondo C, et al. Impact of HMG-CoA reductase inhibition on small GTPases in the heart. *Cardiovasc Dis* 2002; 53: 911-20.
[126] Takemoto M, Node K, Nakagami H, et al. Statins as anti-oxidant therapy for preventing cardiac myocyte hypertrophy. *J Clin Invest* 2001; 108: 1429-37.
[127] Node K, Fujita M, Kitakaze M, et al. Short-term statin therapy improves cardiac function and symptoms in patients with idiopathic dilated cardiomyopathy. *Circulation* 2003; 108: 839-43.
[128] Bellosta S, Via D, Canavesi M, Pfister P, Fumagalli R, Paoletti R, Bernini F. HMG-CoA reductase inhibitors reduce MMP-9 secretion by macrophages. *Arterioscler Thromb Vasc Biol.* 1998; 18: 1671-8.

In: Statins: Indications and Uses, Safety and Modes of Action ISBN 978-1-60692-103-6
Editor: Gunnar N. Holmquist

Chapter 2

SIDE EFFECTS OF STATINS IN MONOTHERAPY

Helmut Sinzinger[1,2] ***and Bernhard A. Peskar***[3]

[1]Wilhelm Auerswald Atherosclerosis Research Group Vienna, Austria

[2]LIPIDUCATE, Unit for Atherosclerosis and Lipid Disorders, Danubian University Krems, Austria;

[3]Department of Experimental and Clinical Pharmacology, University of Graz, Austria

ABSTRACT

Statins have been shown to be remarkably potent in reducing cardiovascular events and improving patient's survival. Overall, the statins show a rather good safety profile. Apart from reports of rare life threatening side effects, mainly from rhabdomyolysis (particularly in combination treatment with fibrates), knowledge and understanding on a great variety of different side effects occurring quite frequently is limited until now. This review on side effects in statin monotherapy attempts to analyse whether there are predisposing factors or specific properties of one or the other of the compounds concerning the development of side effects such as hepatic and renal function, erectile dysfunction, muscular problems and others. Types, prevalence and association with concurrent diseases are described. Underlying biochemical aspects, such as oxidation injury and eventual therapeutic interventions, are discussed. It seems that the rare severe statin side effects can completely be avoided by adhering to 5 simple rules, while the mild side effects probably are much more prevalent than considered so far.

INTRODUCTION

Statins are the number one family of drugs for the treatment of hyperlipidemia taken by more than 100 million people worldwide in these days. They meanwhile belong to the most widely used drugs in medicine. They are acting via inhibition of 3-hydroxy-3-methylglutaryl-coenzyme A reductase (HMG-CoA-reductase), the rate limiting enzyme of cholesterol biosynthesis inducing LDL-receptor upregulation and consequently enhancing cholesterol clearance from blood. Before the introduction of statins lens opacities and oncogenicity were a major concern. However, the statins have been proven remarkably safe in clinical routine. In contrast to earlier findings in in-vitro [111] and experimental studies, even a tumoricidic action alone [40] as well as in combined therapy [29,112] has been discovered. In addition, statins have been shown to limit the initial steps in virus replication [25], relevant for HIV-infection therapy. In general, statins improve patients survival and reduce vascular events with only minimal toxicity most likely via lipid and non-lipid actions.

The Problem

Statins recently attracted increasing pharmacological interest for their wide range of non-lipid (pleiotropic) effects as well as the great variety of side effects most of which are still not yet well understood. In general, this family of compounds is well tolerated and adverse events are generally mild such as flatulence and gastrointestinal discomfort. The incidence of more serious side effects was reported to range between 1 and 7% [18,19]. A review [9] of the available statins in the UK (pravastatin, simvastatin, atorvastatin and rosuvastatin) revealed a comparable rate of adverse events for the 4 compounds resulting in drug withdrawal of about 3% (2.5 – 3.2).

Initially, the large studies involving ten thousands of patients [8,80,83,98] taking different statins failed to demonstrate a difference in muscular side effects between verum and placebo, the actual rates of myotoxicity ranging below 1%. In 1999 we for the first time described exercise-induced muscle pain without CK-elevation [84]. A review [100] up to 2002 found that reports of muscle problems during statin clinical trails are extremely rare.

Phillips et al. [68] confirmed the occurrence of myopathy with normal CK in association with statin therapy in muscle biopsy samples. Fatal rhabdomyolysis is extremely rare. 31 cases after 9.8 million prescriptions have been reported in the United States [94]. Meanwhile, after withdrawal of cerivastatin and due to more careful prescription, this figure has even further improved. Rhabdomyolysis was less than one in 1 000 000, the incidence being comparable among all statins. Elevation of CK to more than 10 times normal occurs in 1 out of 10 thousand patients/year on statin use only. While the incidence of serious muscle problems is very low, the rate of mild side effects has been heavily underestimated so far.

Types of Adverse Events

1. Liver Function

Studies in animals and early data in human revealed hepatotoxicity, mainly minor elevation of alanine aminotransferase (ALT). ALT levels more than three times the upper normal level were reported at 2.6% at low (20 mg/d) and 5.0% at high (80 mg/d) doses of lovastatin [101]. The recently finished Treating to new targets (TNT)-study showed an increase in transaminases after atorvastatin 10 mg in 0.2% and at the 80 mg dose of 1.2%

eventually indicating dose-dependency. The prevalence seems to be similar for other statins. There was, however, no difference in liver enzyme elevations when patients were grouped according to LDL achieved (< 40, 41 – 60, 61 – 80, 81 – 100 mg/dl with a rate of 2.6, 3.2, 3.0 and 3.1, respectively) (PROVE-IT). Acute hepatic failure (< 1 : one million patients treatment years) is close to the background rate. Pretherapeutically elevated liver enzymes do not seem to increase the rate of liver enzyme increase nor does hepatopathy secondary to alcohol abuse. Interestingly, patients with a positive history for hepatitis A have a significantly worse response concerning liver enzymes, but not with hepatitis B and C in anamnesis. Discontinuation of statin therapy results in rapid normalization of hepatic enzymes. Interestingly, for unknown reasons, in some patients enzymes may normalize despite ongoing therapy at unchanged statin dose. In combination therapy with fenofibrate ALT increase > 3 times the upper limit of normal did not differ to monotherapy, while an increased rate of milder elevation occurred in such combination therapy [33].

2. Renal Function

Hyperlipidemia plays an active role in the progression of renal disease. Statins have been reported to partially restore renal function. Recently we found that atorvastatin improves creatinine clearance as well as microalbuminuria (MAU) in hypercholesterolemic patients [90], the benefit not differing between a 10 and 40 mg daily dose. The maximum benefit was achieved at 3 months, plateauing thereafter. Interestingly, in some patients at high doses (simvastatin ≥ 80 mg, atorvastatin ≥ 80 mg, rosuvastatin ≥ 40 mg) MAU may exhibit a significant increase. The basis of this J-shape response is unknown. A contrasting view has been presented [106] that mild proteinuria occurring on all statins is generally transient and reversible and not associated with deterioration in renal function. A change in endocytosis has been claimed for the tubular proteinemia [105]. Whether this view allows to continue statin therapy despite impairment in the patients affected is questionable.

3. Erectile Dysfunction

Dyslipidemia favours development and progression of atherosclerotic vascular disease which again is strongly linked to erectile dysfunction (ED) [76]. Dyslipidemia is known to be increasingly linked with ED [79,108]. Lipid lowering by statins improves ED [78]. Since long it has been claimed that the intake of hypolipidaemic drugs [10] is associated with an increased rate of ED. A literature review in 2002 [72] revealed that a substantial number of cases of ED associated with statin intake have been reported to regulatory agencies, supporting the suggestion that statins also cause ED. As patients rarely report spontaneously on ED [46] it is not appropriate to conclude that statins are unlikely to cause impotence [66]. Spontaneous reporting severely underestimates ED. Using a standardized validated questionnaire [73], however, we found [110] a clear dose-dependent increase in ED on atorvastatin at 20 mg and even more at a 40 mg daily dose. ED was not related to other side effects (CK, liver enzymes, muscular complaints, etc). Monitoring the intake of other statins and its possible correlation with ED is under way at present. The pathomechanism is not clear yet. After statin discontinuation complete normalization of erectile function may take weeks although so far no controlled data are available.

4. Muscular Problems

In middle aged people bone, joint and muscle symptoms are not uncommon. Muscular complaints have been reported in clinical trials of statins, but surprisingly to a comparable frequency and severity in verum and placebo groups. This may be one reason why these symptoms have been neglected for such a long time. Usually, if a patient takes a drug eventually creating muscle problems, symptoms are rapidly attributed to this medication. Surprisingly not so with statins. Distributing a validated questionnaire revealed that patients are tending to underreport statin induced muscle symptoms. Muscular complaints are quite common among patients on statin therapy, in the great majority with normal creatine kinase. No anatomic map as to the location and involvement of muscle groups has been provided so far although proximal muscles seem to be involved more frequently. The likelihood that symptoms appear on statins is decreasing with treatment time in general, depending on the type of manifestation. Usually clinical symptoms disappear almost immediately after statin withdrawal, however, muscular weakness may persist – although at lower intensity – for up to 1 year. In contrast to an advisory statement [65] and an editorial comment [32] we saw no correlation to age, sex and body weight in more than 500 patients with muscular complaints. Interestingly, most of these patients were on monotherapy. No significant difference for the statins currently in use has been shown so far concerning prevalence or any special type of muscular side effects. Patients may tolerate one statin, but not another one, for unknown reasons [86]. Therefore, switching should be considered first rather than discontinuing [32]. Only very rarely patients do not tolerate any of the available statins. Earlier terminology on muscle symptoms was confusing. Nowadays, muscular adverse events usually are classified into:

- Myalgia – muscle pains
  CK normal or elevated
  dose-dependency not clarified
- Myopathy – muscle pains plus CK elevation
  positive EMG
  dose-dependency not clarified
- Myositis – inflammatory component
  CK – normal or elevated; CRP-normal or elevated
  dose-dependency not clarified
- Rhabdomyolysis – muscle destruction, myoglobin release, muscle pain;
  dose-dependent

CK is always increased; may cause renal failure and in certain cases turn out to be fatal. All the described events are reversible if diagnosed in time (e.g. rhabdomyolysis).

Up to a dose of 0.4 mg, CK-increase on cerivastatin does not appear to be different from other statins. The majority of cases of rhabdomyolysis with cerivastatin, however, was seen at a dose of 0.8 mg, particularly in combination with fibrates (gemfibrozil).

Various types of symptoms can be differentiated:

- ache-like symptoms – usually appearing immediately after exercise, rather promptly disappearing
- cramp-like symptoms – the onset in the majority of patients appears on the day after exercise
- burning sensation – mainly after exercise, onset immediately
- flu-like symptoms – with acute heavy muscle pain [87] and oppressive pain are the most severe and acute forms. They are accompanied by an increase in inflammatory markers (CRP, interleukin 6, etc.), fever and often arthralgia
- weakness – tonometrically proven, usually appears rather late, even months or years after initiation of statin therapy
- myositis migrans – muscle symptoms, mainly ache-like moving over certain muscles in a characteristic individual sequence [88]

In an editorial [32] to a paper of Phillips [68] Grundy discusses the question whether statin can cause chronic low-grade myopathy. Small size, random variation and lack of appropriate controls in the study [68] are critically discussed.

From the fact that controlled studies did not detect a higher prevalence of muscle symptoms during statin therapy it was concluded that statin-associated myopathy with normal CK levels essentially does not exist or that, if it does exist, it cannot be detected above a background of muscle symptoms in the general clinical trial population. As the clinical picture is so clear and the prevalence definitely several-fold higher than mentioned in the large trials, an experienced physician regularly seeing these patients hardly can share and understand this view. The question that the methodology to assess the symptoms in these large trials might have been insufficient, however, has so far not been asked.

Addressing the benefit of statins the editorial [32] claims that before discontinuing therapy, physicians should carefully evaluate any patient receiving statins who reports muscle symptoms. The possibility to switch to another statin is not even touched, although that approach is successful in > 75% of such patients.

A data base consisting of 972 myopathy surveys [41] revealed that 36% showed cramps at night, 20% cramps after exercise, 87% aches or soreness, 67% fatigue or tiredness and 59% weakness. Prevalence was comparable between females and males. A survey of 81 rhabdomyolysis patients [41] revealed, that 51% were never regaining full muscular strength.

Statin-associated muscular problems are considered to be the most frequent and severe side effect of this class of drugs [21,34], its prevalence in large, controlled, prospective studies being about 5% (1 – 7% [17,19]). In a review doing a PubMed search up to January 2003, the frequency of less serious muscle events such as pains and weakness is claimed to range between 1 and 5% [100]. In the recently presented TNT study myalgia prevalence did not differ between the low-dose (10 mg atorvastatin) and the high-dose (80 mg) group (4.7

vs. 4.8%). There is no doubt, however, that in daily clinical routine the prevalence is substantially higher, the severity, however, widely varying.

Mechanisms:

Rate of myotoxicity is comparable between the statins that are metabolised via the cytochrome system and the ones that are not. It appears to be a class effect. The claim that lipophilic statins may be more myotoxic due to the higher penetration of the myocyte [20] could not be substantiated. Although there seems to be some dose-relation, a clear dose-dependency still needs to be assessed for the various forms of myotoxicity. When relating myositis as well as CK-elevation to the LDL-achieved (4 groups 100 to 40 mg/dl) the side effects did not exhibit any difference between the groups (PROVE-IT).

A variety of pathogenetic hypotheses such as disturbed interruption of glycoprotein synthesis in the myocyte membrane [93], chloride channel activation deficiency [20], impaired sodium-potassium channel function [48], increased intracellular calcium concentration [36] and altered membrane fluidity due to reduced cholesterol [55] have been discussed. Statin induced apoptosis mediated via depletion of geranyl geranylated proteins is an even more likely explanation. Even substantial apoptosis often may histologically appear normal.

Phillips et al. claimed that statin-associated muscle problems might result from impaired fatty acid oxidation [69]. In more than one thousand patients with muscle complaints on statins a mean triglycerides level of 341 mg/dl was observed, indicating a higher prevalence in hypertriglycidemia [41].

Differentiation from non-statin causes such as aging, degenerative skeletal disorders, aches, etc. is difficult to almost impossible. Severity may range from rhabdomyolysis and polymyositis to increasingly frequent but milder forms such as myalgia (6.2 – 9.1% prevalence according to WHO), ache- or cramp-like symptoms, weakness, and tiredness. EMG and other imaging modalities [50] have diagnostically failed.

It has been stated that myopathy in association with statins is dose-related concerning CK-increase. Looking at the data, especially for atorvastatin and rosuvastatin, the dose-relationship is unclear and in many cases even doubtful.

Muscle biopsies identified mitochondrial dysfunction. The IMPOSTER trial, a double-blind, randomised cross-over designed study by the group of Phillips, showed that two thirds of patients correctly identified statin therapy underlying their muscle symptoms. In this study all of those that were biopsied showed histological signs of mitochondrial dysfunction.

The role of exercise:

Thompson and coworkers were the first to describe exercise-induced muscle injury with CK-elevation but in absence of clinical symptoms after lovastatin [99]. In recent reviews [62,103] this problem has not been discussed. In a double-blind, placebo-controlled crossover study of 10 healthy young men undergoing controlled exercise in 1991 [71] the authors suggested that lovastatin is not an independent risk factor for developing exercise-induced muscle damage. A couple of years later, Smit et al. [92] reported that the exercise-induced increase in myoglobin- and CK-levels did not differ between untreated and simvastatin treated FH-patients. If regular exercise is performed, these symptoms on statins become more prevalent and severe. Top athletes have been shown to tolerate only rarely statins [91].

Oxidation:

a) LDL-oxidation:

In fact, most data have shown a decreased susceptibility of LDL to oxidation in-vitro as well as ex-vivo [43,74,104] using a variety of different tests.

One of the few studies showing increased LDL-susceptibility to oxidation during statin administration found only partial restoration of antioxidant capacity of LDL when CoQ10 was coadministered [38].

b) Isoprostanes:

Oxidation injury at present is one major concern in patients on statins. The isoprostane 8-epi-$PGF_{2\alpha}$ has been shown to be a reliable marker for in-vivo oxidation injury [56]. Interestingly, normally statins are associated with a decrease in oxidation injury (decrease in isoprostane 8-epi-$PGF_{2\alpha}$ [62]) and reduced oxidizability of lipoproteins (LDL, HDL, VLDL). A decrease in 8-epi-$PGF_{2\alpha}$ has been shown for various statins in experimental animals and patients as well [47,62]. Rhabdomyolysis has been shown to be linked to increased lipid peroxidation [37]. Oxidation injury at the mitochondrial level may be the underlying pathomechanism. A certain not yet defined group may react with an increase in 8-epi-$PGF_{2\alpha}$ for unknown reasons. This increase shows no correlation to either CK-increase or muscular symptoms [61]. The onset of this increase may be as late as 2 years after initiation of therapy, although in the overwhelming majority it is seen as soon as within a few weeks only.

5. Arthritis

Pleiotropic actions of statins may interfere with acute inflammatory processes as experimental and in-vitro data indicate [67]. Statins have also been found to be of clinical benefit in rheumatoid arthritis patients [1,54]. A marked suppression of acute phase reactants was paralleled by a modest improvement in disease activity after a 6-months treatment with atorvastatin [53]. In contrast, however, some patients may develop arthritis symptoms affecting mainly the small finger joints. They seem to be promptly reversible after statin withdrawal. Sometimes, arthralgia is associated with the flu-like response as a secondary event. In a typical patient we were even able to document improvement by means of bone scintigraphy after stopping atorvastatin (10 mg/d) therapy. Examining 5674 elderly women, Beattie et al. [4] found at an 8 year follow-up an increased (OR 1.92) risk of developing incident hip osteoarthritis. In contrast, however, there was a consistent non-significant trend towards decreased progression in statin users. Controlled, prospective studies to assess the effect of statins on the disease activity in chronic inflammatory disease and rheumatoid arthritis are not available.

6. Lupus-Like Syndrome

It has been shown [60] that statins may induce a systemic autoimmune reaction. The statin-induced lupus-like syndrome shows a long delay between initiation of treatment and skin eruption. Despite discontinuation of the drug, antinuclear antibodies may persist for many months. A patient who in addition developed severe autoimmune hepatitis [30] during treatment with atorvastatin has been described. The 3 patients with skin manifestations we saw (1 atorvastatin, 1 simvastatin, 1 pravastatin) had symptoms after stopping statins for further 4 – 11 months. Another patient with simvastatin induced diffuse interstitial pneumonia [45] has been reported. Many authors conclude [45,60] that both lupus-like syndrome as well as interstitial lung disorders may occur more frequently, with many cases remaining unrecognised. Early diagnosis would be warranted and the causal relationship is difficult to establish.

7. Tendinopathy

So far 4 cases [12] of tendinopathy in patients receiving statin therapy have been reported (2 Achilles, 1 tibialis anterior, 1 hand), 2 of them being on simvastatin, 2 on atorvastatin. Symptoms manifested 1 – 2 months after initiation of treatment and disappeared 1 – 2 months after discontinuation. Analysing our patients we found a surprisingly high number (n = 11) of Achilles tendon ruptures in patients below the age of 50 years. It is, however, difficult to assess the relation of tendinopathy with statin use. Eventual differences concerning the different compounds, dose or duration of treatment have not been studied yet. Prospective sonographic monitoring of the Achilles tendon from initiation of statin treatment could help to clarify this issue.

8. Polyneuropathy

The possible association between statin use and polyneuropathy has been reviewed by 2 independent groups [3,13], both suggesting that a risk may exist, prevalence and intensity [22] being low and being small as compared to the benefits. Among a group of case reports [49,82,109] one case resembling Guillain-Barre syndrome [70], seems particularly interesting. In all patients symptoms disappeared latest within 1 year after statin withdrawal [39].

A higher incidence in peripheral neuropathy (1/14000 on statins a year) has been reported in a population-based dynamic cohort study. Thick and thin nerve fibres seem to be equally affected. Despite statin withdrawal, neuropathy may persist for a year(s). Data on dose, type of statin, and duration of treatment, pathomechanism as well as differentiation from other forms of polyneuropathy are not yet available. Statin-associated exacerbation of myasthenia has been reported [11]. Severe central nervous system anomalies have been reported in first pregnancy in trimester statin exposure [15].

9. Cognitive Function

Some statins have been reported to cause impairment of daytime cognitive processes. Experimental data in rodents indicate an impaired learning process with statins [23,42,75]. A review on 60 cases of statin-associated memory loss, depression, sleeping disorders and global amnesia supported that claim [107]. In a double-blind randomised placebo-controlled study (n = 209; duration 6 months) with lovastatin detrimental effects on cognitive performance (attention, working memory, mental efficiency) have been described [58]. The same group found a minor decrement in cognitive function also for simvastatin using a similar study design [59]. In contrast, however, a large body of investigations was not able to substantiate that claim [5,16,23,35,42,75,113], among professions tested such as pilots and aircrew [24]. A retrospective cohort study found that statins are ameliorating cognitive decline in older people [95]. Impaired cognitive function, dementia or depression [14], in turn may result in a very low adherence to medication regimen [96].

The ongoing UCSD Statin Study enrolling 1000 patients to take part in a randomized placebo-controlled trial simvastatin, pravastatin and placebo for 6 months and a 2-months postcessation follow-up for the first time prospectively assesses adverse issues such as cognition, behaviour measure depression, quality of life, mood, sleep and secondary aggression measures [27].

10. Depression/Suicide

A case report discusses the relation of statin therapy and depression [52]. Examining long-term statin use was associated with a lower risk of abnormal depression scores (OR 0.63), anxiety (OR 0.69) and hostility (OR 0.77) [115]. A review found that in recent statin trials deaths from accidents and suicide were slightly fewer with statins as compared to placebo [51]. Analysing the United Kingdom General Practice Research Data Base (1991-1999) discovered no increased risk of depression or suicide in patients on statins [114]. Case reports were also published on aggression [28]. This possible effect is now assessed in a controlled trial [27].

THERAPEUTIC INTERVENTION

Coenzyme Q10

CoQ10 is an important redox component [6]. Inhibition of mevalonate formation and derivates such as coenzyme Q10 (ubiquinone) is one of the aspects, which eventually could have a negative effect decreasing antioxidative potential. In fact, it has consistently been shown that CoQ10 is lowered during statin therapy [7,31,44,77]. The decline was found to be dose-dependent for lovastatin and pravastatin [57]. Ubiquinone is used by mitochondria for electron transport, a lower level has been observed in some mitochondrial myopathies. Some authors reported comparable light microscopic findings in biopsy samples [26,81] from statin treated patients. The resulting defective ATP synthesis at the mitochondrial level might contribute eventually to cellular instability. A decrease in CoQ10 usually ranging between 10 and 15% is accompanied by a LDL-decrease of almost always more than 20% even if the lowest available statin dose is used. After stopping statin therapy, after a temporary rebound increase of CoQ10 by sometimes more than 30% can be monitored. Vitamin E and CoQ10, important antioxidants for LDL, the one more abundant, the other more potent [64,97] are carried among others (VLDL, HDL) by this lipoprotein. Apparently, LDL is lowered by statin therapy more severely as CoQ10 and Vitamin E for example. Calculating therefore the CoQ10/LDL-ratio and vitamin E/LDL-ratio in almost all patients there is an increase rather than a decrease. An attempt in treating patients with muscular symptoms during statin administration concomitantly with CoQ10 did not improve or change the symptoms in any of the 21 patients (12 males, 9 females; 36 – 57 years) throughout a 3-months period (own unpublished results). During lovastatin administration enhanced oxidizability of ubiquinol and α-tocopherol was found [63]. Ubiquinone supplementation also was not associated with a significant improvement of antioxidant capacity [64]. Although no data are available to support this conclusion, the ACCC/AHA NHLBI clinical advisory claims a coenzyme Q10-deficiency as a possible mechanism [65]. There is only 1 patient report demonstrating improvement of muscular symptoms if vitamin E was coadministered to statins, while in large numbers of patients this approach was not successful [85].

L-Carnitine

Toma claimed that L-carnitine might exert a favourable action on statin-induced myotoxicity [102]. Lovastatin administration in rabbits resulted in a decreased tissue level of carnitine [2]. Experimental studies on simvastatin treated at a myotoxic dose (70 – 210 mg/kg) rats with coadministration of carnitine (200 mg/kg b.i.d.) revealed counteraction of CK-elevation and improved treadmill testing, but no significant effect on elevated hepatic enzymes. Preliminary data on carnitine administration in patients showing muscular side effects on various statins, however, revealed no benefit either on symptoms or elevated CK. Symptom-free elevation of 8-epi-$PGF_{2\alpha}$ was not beneficially affected as well. The acylcarnitine-free carnitine ratio failed to predict and identify patients with muscle complaints. Thus, no benefit has been proven on any of the known statin-induced adverse events.

DISCUSSION

The US Statin Advisory recommends continuing statin therapy despite muscle complaints, if CK levels do not exceed 10 times the upper limit of normal value. This view is in sharp contrast to the one in Europe. CK is definitely an inadequate parameter to exclude muscle toxicity of statins. Actually, no parameter tested so far turned out to be clinically useful. Muscular symptoms are reproducible in the overwhelming majority of patients at repeated reexposure to statins. They usually occur despite normal CK, while interestingly, frequently patients responding with CK-elevation do not appear to show symptoms, at least when CK is elevated only moderately.

Although it is very clear from the clinical point of view that statins are inducing different types of muscle complaints, assessment becomes difficult. Disappearance of symptoms after withdrawal of the respective statin or correct identification of statins and placebo by the patients may be insufficient [68]. The true rate of occurrence is unclear. Large trials definitely underestimate most of the adverse events reported in this review. Long delay in onset mostly very mild forms of manifestation and eventually delayed disappearance of problems does not allow a cross-over study design.

A prospective blinded study focused to assess the exact rate of adverse events with even only one statin would require an extremely high number of patients (above 10,000) because of the difficulty to differentiate mild side effects from the background. Otherwise, adverse event rate might be underestimated again. The first attempt in this direction with about 1,000 patients is currently performed [27].

The claim to use serum aldolase to identify patients experiencing CK-negative statin myopathy [38] cannot be proven. Whenever tested, aldolase values in myopathy patients were normal.

Another parameter, which could reflect destroyed cellular membranes, myoglobin, failed to identify patients with either symptoms and/or CK-elevation.

Conclusion

Statins remarkably increased the therapeutic options in prevention of vascular disease and effects beyond. Although remarkably safe individual selecting, dosing and monitoring is required. In order to minimize adverse events the 5 golden rules of statin therapy [89] for patients have been reported:

1. Start always with the lowest available dose for at least 4 weeks.
2. Avoid combination with fibrates.
3. At the onset of muscular symptoms stop statin treatment immediately even before contacting your physician.
4. When taking a new drug ask your physician before whether this combination is compatible. If this is not possible and the new drug is important, the statin-treatment should be discontinued until clarification.
5. If (macrolid-)antibiotics or antimycotics have to be taken, statin-therapy has to be discontinued for the entire duration of the antibiotic/antimycotic treatment.

Acknowledgements

We are indebted for the helpful comments of Karl Zeiler, MD, Prof. (Department of Neurology, Medical University of Vienna) and the valuable help in preparing and critically reading the manuscript by Eva Unger und Susanne Granegger.

REFERENCES

[1] Abud-Mendoza, C; de la Fuenta, H; Cuevas-Orta, E; Baranda, L; Cruz-Rizo, J; Gonzalez-Amaro, R. (2003). Therapy with statins in patients with refractory rheumatic diseases: a preliminary study. *Lupus, 12,* 607-611.

[2] Arduini, A; Peschechera, A; Giannessi, F; Carminati, P. (2004). Improvement of statin-associated myotoxicity by l-carnitine. *J Thromb Haemost, 2*, 2270-2271.

[3] Backes, JM; Howard, PA. (2003). Association of HMG-CoA reductase inhibitors with neuropathy. *Ann Pharmacother, 37,* 274-278.

[4] Beattie, MS; Lane, NE, Hung, YY; Nevitt, MC. (2005). Association of statin use and development and progression of hip osteoarthritis in elderly women. *J Rheumatol,* 32, 105-110.

[5] Berk-Planken, I; deKonig, I; Stolk, R; Jansen, H; Hoogerbrugge, N. (2002). Atorvastatin, diabetic dyslipidemia, and cognitive functioning. *Diabetes Care, 25,* 1250-1251.

[6] Beyer, RE; Nordenbrand, K; Ernster, L. (1987). The function of coenzyme Q in free radical production and as an antioxidant: a review. *Chem Scripta, 27,* 145-153.

[7] Bhuiyan, J; Seccombo, DW. (1996). The effects of 3-hydroxy-3-methylglutaryl-CoA reductase inhibition on tissue levels of carnitine and carnitine acyltransferase activity in the rabbit. *Lipids, 31,* 867-870.

[8] Bradford, RH; Shear, CL; Chremos, AN; Dujovne, C; Downton, M; Franklin, FA; Gould, AL; Hesney, N; Higgins, J; Hurley, DP. (1991). Expanded clinical evaluation of lovastatin (EXCEL) study results. Efficacy in modifying plasma lipoproteins and adverse event profile in 8245 patients with moderate hypercholesterolemia. *Arch Intern Med, 151,* 25-28.

[9] Brewer, HB. (2003). Benefit-risk assessment of rosuvastatin 10 to 40 milligrams. *Am J Cardiol, 92,* 23K-29K.

[10] Bruckert, E; Giral, P; Heshmati, HM; Turpin, G. (1996). Men treated with hypolipidaemic drugs complain more frequently of erectile dysfunction. *J Clin Pharm Ther, 21,* 89-94.

[11] Cartwright, MS, Jeffery, DR; Nuss, GR; Donofsio, PD. (2004). Statin associated exacerbation of myasthenia gravis. *Neurology, 62,* 2188.

[12] Chazerain, P; Hayem, G; Hamza, S; Best, C; Ziza, JM. (2001). Four cases of tendinopathy in patients on statin therapy. *Joint Bone Spine, 68,* 430-433.

[13] Chong, PH; Boskovich, A; Stevkovic, N; Bartt, RE. (2004). Statin-associated peripheral neuropathy: review of the literature. *Pharmacotherapy, 24,* 1194-1203.

[14] Doser, S; Marz, W; Reinecke, MF; Ringleb, P; Schultz, A; Schwandt, P; Becker, HJ; Bonner, G; Buerke, M; Diener, HC; Gohlke, H; Keil, U; Ringelstein, EB; Steinmetz, A; Gladisch, R; Wehling, M. (2004). Recommendations for statin therapy in the elderly. *Internist, 45,* 1053-1062.

[15] Edison, RJ; Muenke, M. (2004). Central nervous system and limb anomalies in case reports of first-trimester statin exposure. *New Engl J Med, 350,* 1579-1582.

[16] Etminan, M; Gill, S; Samii, A. (2003). The role of lipid-lowering drugs in cognitive function: a meta-analysis of observational studies. *Pharmacotherapy, 23,* 726-730.

[17] Evans, M; Rees, A. (2002). Effects of HMG-CoA reductase inhibitors on skeletal muscle - are all statins the same? *Drug Safety, 25,* 649-663.

[18] Evans, M; Rees, A. (2002). The myotoxicity of statins. *Current Opinion in Lipidology, 13,* 415-420.

[19] Farmer, JA; Torre-Amione, G. (2000). Comparative tolerability of the HMG-CoA reductase inhibitors. *Drug Safety, 23,* 197-213.

[20] Gadbut, AP; Caruso, AP; Gabler, JB. (1995). Differential sensitivity of C2-C12 striated muscle cells to lovastatin and pravastatin. *J Mol Cell Cardiol, 27,* 1191-1199.

[21] Gaist, D; Garcia Rodriguez, LA; Huerta, C; Hallas, J; Sindrup, SH. (2001). Lipid lowering drugs and risk of myopathy: a population-based follow-up study. *Epidemiology, 12*, 565-569.

[22] Gaist, D; Garcia Rodriguez, LA; Huerta, C; Hallas, J; Sindrup, SH. (2001). Are users of lipid-lowering drugs at increased risk of peripheral neuropathy? *Eur J Clin Pharmacol, 56*, 931-933.

[23] Gengo, F; Cwudzinski, D; Kinkel, P; Block, G; Stauffer, L; Lines, C. (1995). Effects of treatment with lovastatin and pravastatin on daytime cognitive performance. *Clin Cardiol, 18,* 209-214.

[24] Gibellato, MG; Moore, JL; Selby, K; Bower, EA. (2001). Effects of lovastatin and pravastatin on cognitive function in military aircrew. *Aviat Space Environ Med, 72,* 805-812.

[25] Giguere, JF; Tremblay, MJ. (2004). Statin compounds reduce human immunodeficiency virus type 1 replication by preventing the interaction between virion-associated host intercellular adhesion molecule 1 and its natural cell surface ligand LFA-1. *J Virol, 78,* 12062-12065.

[26] Giordano, N; Senesi, M; Mattii, G; Battisti, E; Villanova, M; Gennari, C. (1997). Polymyositis associated with simvastatin. *Lancet, 349,* 1600-1601.

[27] Golomb, BA; Criqui, MH; White, HL; Dimsdale, JE. (2004). The UCSD Statin Study: a randomised controlled trial assessing the impact of statins on selected noncardiac outcomes. *Control Clin Trials, 25,* 178-202.

[28] Golomb, BA; Kane, T; Dimsdale, JE. (2004). Severe irritability associated with statin cholesterol-lowering drugs. *Quart J Med, 97,* 229-235.

[29] Graaf, MR; Richel, DJ; van Noorden, CJ; Guchelaar, HJ. (2004). Effects of statins and farnesyltransferase inhibitors on the development and progression of cancer. *Cancer Treat Rev, 30,* 609-641.

[30] Graziadei, IW; Obermoser, GE; Sepp, NT; Erhart, KH; Vogel, W. (2003). Drug-induced lupus-like syndrome associated with severe autoimmune hepatitis. *Lupus, 12,* 409-412.

[31] Grinlanda, G; Oradei, A; Manto, A; Lipaa, S; Uccioli, L; Caputo, S; Greco, AV; Littaru, GP. (1993). Evidence of plasma CoQ10-lowering effect by HMG-CoA-reductase inhibitors: a double-blind, placebo-controlled study. *J Clin Pharmacol, 33,* 226-229.

[32] Grundy, SM. (2002). Can statins cause chronic low-grade myopathy? *Ann Intern Med, 137,* 617-618.

[33] Grundy, SM; Vega, GL; Yuan, Z; Battisti, WP; Brady, WE; Palmisano, J. (2005). Effectiveness and tolerability of simvastatin plus fenofibrate for combined hyperlipidemia. The Safari Trial. *Am J Cardiol, 95,* 462-468.

[34] Hamilton-Craig, I. (2001). Statin-associated myopathy. *Med J Aust, 175,* 486-489.

[35] Harrison, RW; Ashton, CH. (1994). Do cholesterol-lowering agents affect brain activity? A comparison of simvastatin, pravastatin, and placebo in healthy volunteers. *Br J Clin Pharmacol, 37*, 231-236.

[36] Hochgraf, E; Levy, Y; Aviram, M; Brook, JG; Cogan, U. (1994). Lovastatin decreases plasma and platelet cholesterol levels and normalizes elevated platelet fluidity and aggregation in hypercholesterolemic patients. *Metabolism, 43,* 59-64.

[37] Holt, S; Reeder, B; Wilson, M; Harvey, S; Morrow, JD; Roberts, LJII; Moore, K. (1999). Increased lipid peroxidation in patients with rhabdomyolysis. *Lancet, 353,* 358-359.

[38] Hyman, MH. (2003). Statin-associated myopathy with normal creatine kinase levels. *Ann of Int Med, 138,* 1007.

[39] Jeppesen, U; Gaist, D; Smith, T; Sindrup, SH. (1999). Statins and peripheral neuropathy. *Eur J Clin Pharmacol, 54,* 835-838.

[40] Keyomarsi, K; Sandoval, L; Band, V; Pardee, AB. (1991). Synchronization of tumor and normal cells from G1 to multiple cell cycles by lovastatin. *Cancer Res, 51,* 3602-3609.

[41] Kordas, KC; Phillips, PS; Golomb, BA. (2004). Clinical characteristics of 1053 patients with statin associated muscle complaints. *Arterioscler Thromb Vasc Biol, 24,* e-133.

[42] Kostis, JB; Rosen, RC; Wilson, AC. (1994). Central nervous system effects of HMG CoA reductase inhibitors: lovastatin and pravastatin on sleep and cognitive performance in patients with hypercholesterolemia. *J Clin Pharmacol, 34,* 989-996.

[43] Kural, BV; Orem, C; Uydu, HA, Alver, A; Orem, A. (2004). The effects of lipid-lowering therapy on paraoxonase activities and their relationships with the oxidant-antioxidant system in patients with dyslipidemia. *Coron Artery Dis, 15,* 277-283.

[44] Laaksonen, R; Ojala, J-P; Tikkanen, MJ; Himberg, J-J. (1994). Serum ubiquinone concentrations after short- and long-term treatment with HMG-CoA reductase inhibitors. *Eur J Clin Pharmacol, 46,* 313-317.

[45] Lantuejoul, S; Brambilla, E; Brambilla, C; Devouassoux, G. (2002). Statin-induced fibrotic non-specific interstitial pneumonia. *Eur Respir J, 19,* 577-580.

[46] Levine, LA; Kloner, RA. (2000). Importance of asking questions about erectile dysfunction. *Am J Cardiol, 86,* 1210-1213.

[47] Li, W; Asagami, T; Matsushita, H; Lee, KH; Tsao, PS. (2005). Rosuvastatin attenuates monocytes-endothelial cell interactions and vascular free radical production in hypercholesterolemic mice. *J Pharmacol Exp Ther, 21*, (in publication).

[48] Lijnen, P; Celis, H; Fagard, R; Staessen, J; Amery, A. (1994). Influence of cholesterol lowering on plasma membrane lipids and cationic transport systems. *J Hypertens, 12,* 59-64.

[49] Lo, YL; Leoh, TH; Loh, LM; Tan, CE. (2003). Statin therapy and small fibre neuropathy: a serial electrophysiological study. *J Neurol Sci, 15,* 105-108.

[50] Lupattelli, G; Palumbo, B; Sinzinger, H. (2001). Statin-induced myopathy does not show up in MIBI-scintigraphy. *Nucl Med Commun, 22,* 575-578.

[51] Manfredini, R; Caracciolo, S; Salmi, R; Boari, B; Tomelli, A; Gallerani, M. (2000). The association of low serum cholesterol with depression and suicidal behaviours: new hypothesis for the missing link. *J Int Med Res, 28,* 247-257.

[52] Mayo Clinic Womens Healthsource. (2004). One on one. Several months ago, my doctor prescribed a statin for my high cholesterol. Since that time, I´ve become depressed. Did the statin cause my depression? *Mayo Clinic Womens Healthsource, 8,* 8.

[53] McCarey, DW; McInnes, IB; Madhok, R; Hampson, R; Scherbakov, O; Ford, I; Capell, HA; Sattar, N. (2004). Trial of atorvastatin in rheumatoid arthritis (TARA): double-blind, randomized placebo-controlled trial. *Lancet, 363,* 2015-2021.

[54] McCarey, DW; Sattar, N; McInnes, IB. (2005). Do the pleiotropic effects of statins in the vasculature predict a role in inflammatory diseases? *Arthritis Res Ther, 7,* 55-61.

[55] Morita, I; Sato, I; Ma, L; Murota, S. (1997). Enhancement of membrane fluidity in cholesterol-poor endothelial cells pretreated with simvastatin. *Endothelium, 5,* 107-113.

[56] Morrow, JD; Minton, TA; Badr, KF; Roberts, LJII. (1994). Evidence that the F_2-isoprostrane 8-epi-prostaglandin $F_{2\alpha}$ is formed in-vivo. *Biochim Biophys Acta, 121,* 244-248.

[57] Mortensen, SA; Leth, A; Agner, E; Rohde, M. (1997). Dose-related decrease in serum coenzyme Q10 during treatment with HMG-CoA reductase inhibitors. *Mol Aspects Med, 18,* S137-S144.

[58] Muldoon, MF; Barger, SD; Ryan, CM; Flory, JD; Lehoczky, JP; Matthews, KA; Manuck, SB. (2000). Effects of lovastatin on cognitive function and psychological well-being. *Am J Med, 108,* 538-546.

[59] Muldoon, MF; Ryan, CM; Sereika, SM; Flory, JD; Manuck, SB. (2004). Randomized trial of the effects of simvastatin on cognitive functioning in hypercholesterolemic adults. *Am J Med, 117,* 823-829.

[60] Noel, B; Panizzon, RG. (2004). Lupus-like syndrome associated with statin therapy. *Dermatology, 208,* 276-277.

[61] Oguogho, A; Mehrabi, M; Sinzinger, H. (1999). Increased plasma, serum and urinary 8-epi-prostaglandin $F_{2\alpha}$ in heterozygous hypercholesterolemia. *Wr klin Wschr, 111,* 113-118.

[62] Omar, MA; Wilson, JP; Cox, TS. (2002). Rhabdomyolysis and HMG-CoA reductase inhibitors. *Am Pharmacol Therapy, 35,* 1096-1107.

[63] Palomäki, A; Malminiemi, K; Metsa-Ketela, T. (1997). Enhanced oxidizability of ubiquinol and alpha-tocopherol during lovastatin treatment. *FEBS Let, 410,* 254-258.

[64] Palomaki, A; Malminiemi, K; Solakivi, T; Malminiemi, O. (1998). Ubiquinone supplementation during lovastatin treatment: effect on LDL oxidation ex vivo. *J Lipid Res, 39,* 1430-1437.

[65] Pasternak, RC; Smith, SC; Bairey-Merz, CN; Grundy, SM; Cleeman, JI; Lenfant, C; American College of Cardiology; American Heart Association; National Heart, Lung and Blood Institute. (2002). ACC/AHA, NHLBI clinical advisory on the use and safety of statins. *J Am Coll Cardiol, 40,* 567-572.

[66] Pedersen, TR; Faergeman, O. (1999). Simvastatin seems unlikely to cause impotence. *Brit Med J, 318,* 192.

[67] Penney; CP. (2005). Statins in rheumatology. *J Rheumatol, 32,* 17-19.

[68] Phillips, PS; Haas, RH; Bannykh, S; Hathaway, S; Gray, NL; Kimura, BJ; Vladutiu, GD; England, JD. (2002). Statin-associated myopathy with normal creatine kinase levels. *Ann Int Med, 137,* 581-585.

[69] Phillips, CT; Gray, NL; Puhek, LM; McDonald, FG; Sullivan, MJ; Phillips, PS. (2004). Basal respiratory exchange ratio is altered with statin use in normals. *J Am Coll Card, 43,* 233A.

[70] Rajabally, YA; Varakantam, V; Abbott, RJ. (2004). Disorder resembling Guillain-Barre syndrome on initiation of statin therapy. *Muscle Nerve, 30,* 663-666.

[71] Reust, CS; Curry SC; Guidry, JR. (1991). Lovastatin use and muscle damage in healthy volunteers undergoing eccentric muscle exercise. *West J Med, 154,* 198-200.

[72] Rizvi, K; Hampson, JP; Harvey, JN. (2002). Do lipid-lowering drugs cause erectile dysfunction? A systematic review. *Fam Pract, 19,* 95-98.

[73] Rosen, RC; Cappelleri, JC; Smith, MD; Lipsky, J; Pena, BM. (1999). Development and evaluation of an abridged, 5-item version of the International Index of Erectile Function (IIEF-5) as a diagnostic tool for erectile dysfunction. *Int J Impot Res, 11,* 319-326.

[74] Rosensen, RS; Tangney, CC; Levine, DM; Parker, TS; Gordon, BR. (2005). Association between reduced low-density lipoprotein oxidation and inhibition of monocytes chemoattractant protein-1 production in statin-treated subjects. *J Lab Clin Med, 145,* 83-87.

[75] Roth, T; Richardson, GR; Sullivan, JP; Lee, RM; Merlotti, L; Roehrs, T. (1992). Comparative effects of pravastatin and lovastatin on night-time sleep and daytime performance. *Clin Cardiol, 15,* A26-A32.

[76] Roumeguere, T; Wespes, E; Carpentier, Y; Hoffmann, P; Schulman, CC. (2003). Erectile dysfunction is associated with a high prevalence of hyperlipidemia and coronary heart disease risk. *Eur Urol, 44,* 355-359.

[77] Rundek, T; Naini, A; Sacco, R; Coates, K; DiMauro, S. (2004). Atorvastatin decreases the coenzyme Q10 level in the blood of patients at risk for cardiovascular disease and stroke. *Arch Neurol, 61*, 889-892.

[78] Saltzman, EA; Guay, AT; Jacobson, J. (2004). Improvement in erectile function in men with organic erectile dysfunction by correction of elevated cholesterol levels: a clinical observation. *J Urol, 172,* 255-258.

[79] Schachter, M. (2000). Erectile dysfunction and lipid disorders. *Curr Med Res Opin, 16,* 9-12.

[80] Scandinavian Simvastatin Survival Group. (1994). Randomized trial of cholesterol lowering in 4444 patients with coronary heart disease: the Scandinavian Simvastatin Survival Study (4S). *Lancet, 344,* 1383-1389.

[81] Schalke, BB; Schmidt, B; Toyka, K; Hartung, HP. (1992). Pravastatin associated inflammatory myopathy. *N Engl J Med, 327,* 649-650.

[82] Scola, RH; Trentin, AP; Germiniani, FM; Piovesan, EJ; Werneck, LC. (2004). Simvastatin-induced mononeuropathy multiplex: case report. *Arq Neuropsiquiatr, 62,* 540-542.

[83] Shepherd, J; Cobbe, SM; Ford, I; Isles, CG; Lorimer, AR; MacFarlane, PW; McKillop, JH; Packard, CJ. (1995). Prevention of coronary heart disease with pravastatin in men with hypercholesterolemia. West of Scotland Coronary Prevention Study Group. *N Engl J Med, 333,* 1301-1307.

[84] Sinzinger, H; Schmid, P; O'Grady, J. (1999). Two different types of exercise-induced muscle pain without myopathy and CK-elevation during HMG-Co-enzyme-A-reductase inhibitor treatment. *Atherosclerosis, 143,* 459-460.

[85] Sinzinger, H. (2000). Does vitamin E beneficially affect muscle pains during HMG-Co-enzyme-A-reductase inhibitors without CK-elevation? *Atherosclerosis, 149,* 225.

[86] Sinzinger, H; Wolfram, R; Peskar, BA. (2002). Muscular side effects of statins. *J Cardiovasc Pharmacol, 40,* 163-171.

[87] Sinzinger, H. (2002). Flue-like response on statins. *Med Sci Monit, 8,* 384-388.

[88] Sinzinger, H. (2002). Statin-induced myositis migrans. *Wr klin Wschr, 114,* 943-944.

[89] Sinzinger, H; Laimer, H; Hoppichler, F; Toplak, H; Kritz, H; Kleemann, L; Pilz, H; Grafinger, P; Benzer, W; Brainin, M; Derfler, K; Dzien, A; Pidlich, J; Pirich, C; Lechleitner, M; Ludvik, B; Mayer, G; Silberbauer, K; Traindl, O. (2002). Fünf Goldene Regeln. (Consensus paper). *Österr Ärztezeitung, 12,* 38-40.

[90] Sinzinger, H; Kritz, H; Furberg, CD. (2003). Atorvastatin reduces microalbuminuria in patients with familial hypercholesterolemia and normal glucose tolerance. *Med Sci Monitor, 9,* 188-192.

[91] Sinzinger, H; O'Grady, J. (2004). Professional athletes suffering from hypercholesterolemia rarely tolerate statin treatment because of muscular problems. *Brit J Clin Pharmacol, 57,* 525-528.

[92] Smit, JW; Bar, PR; Geerdink, RA; Erkelens, DW. (1995). Heterozygous familial hypercholesterolemia is associated with pathological exercise-induced leakage of muscle proteins, which is not aggravated by simvastatin therapy. *Eur J Clin Invest, 25,* 79-84.

[93] Sonoda, Y; Gotow, T; Kuriyama, M; Nakahara, K; Arimura, K; Osame, M. (1994). Electrical myotonia of rabbit skeletal muscles by HMG-CoA reductase inhibitors. *Muscle Nerve, 17,* 891-897.

[94] Staffa, JA; Chang, J; Green, L. (2002). Cerivastatin and reports of fatal rhabdomyolysis. *N Engl J Med, 246,* 539-540.

[95] Starr, JM; McGurn, B; Whiteman, M; Pattie, A; Whalley, LJ; Deary, IJ. (2004). Life long changes in cognitive ability are associated with prescribed medications in old age. *Int J Geriatr Psychiatry, 19,* 327-332.

[96] Stilley, CS; Sereika, S; Muldoon, MF; Ryan, CM; Dunbar-Jacob, J. (2004). Psychological and cognitive function: predictors of adherence with cholesterol lowering treatment. *Ann Behav Med, 27,* 117-124.

[97] Stocker, R; Bowry, VW; Frei, B. (1991). Ubiquinol-10 protects human low-density lipoprotein more efficiently against lipid peroxidation than does α-tocopherol. *Proc Natl Acad Sci USA, 88,* 1646-1650.

[98] The Pravastatin Multinational Study Group for Cardiac Risk Patients. (1993). Effects of pravastatin in patients with serum cholesterol levels between 5.2 to 7.8 mmol/l plus two additional atherosclerotic risk factors. *Am J Cardiol, 72,* 1031-1037.

[99] Thompson, PD; Zmuda, JM; Domalik, LJ; Zimet, RJ; Staggers, J; Guyton, JR. (1997). Lovastatin increases exercise-induced skeletal muscle injury. *Metabolism, 46,* 1206-1210.

[100] Thompson, PD; Clarkson, P; Kars, RH. (2003). Statin-associated myopathy. *JAMA, 289,* 1681-1690.

[101] Tolman, KG. (2002). The liver and lovastatin. *Am J Cardiol, 89,* 1374-1380.

[102] Toma, E. (2003). Statin-associated myopathy with normal creatine kinase levels. *Annals of Int Med, 138,* 1007-1008.

[103] Ucar, M; Mjörndal, T; Dahlqvist, R. (2000). HMG-CoA reductase inhibitors and myotoxicity. *Drug Safety, 29,* 441-457.

[104] van den Akker, JM; Bredie, SJ; Diepenveen, SH; van Tits, LJ; Stalenhoef, AF; van Leusen, R. (2003). Atorvastatin and simvastatin in patients on hemodialysis: effects on lipoproteins, C-reactive protein and in-vivo oxidized LDL. *J Nephrol, 16,* 238-244.

[105] Verhulst, A; D'Haese, PC; De Brol, ME. (2004). Inhibitors of HMG-CoA-reductase reduce receptor-mediated endocytosis in human kidney proximal tubular cells. *J Am Soc Nephrol, 14,* 2249-2257.

[106] Vidt, DG; Cressman, MD; Harris, S; Pears, JS; Hutchinson, HG. (2004). Rosuvastatin-induced arrest in progression of renal disease. *Cardiology, 102,* 52-60.

[107] Wagstaff, LR; Mitton, MW; Arvik, BM; Doraiswamy, PM. (2003). Statin-associated memory loss: analysis of 60 case reports and review of the literature. *Pharmacotherapy, 23,* 871-880.

[108] Wei, M; Macera, CA; Davis, DR; Hornung, CA; Nankin, HR; Blair, SN (1994). Total cholesterol and high density lipoprotein cholesterol as important predictors of erectile dysfunction. *Am J Epidemiol, 140,* 930-937.

[109] Weimer, LH. (2003). Medication-induced peripheral neuropathy. *Curr Neurol Neurosci Rep, 3,* 86-92.

[110] Wolfram, RM; Budinsky, A; Kratzik, C; Sinzinger, H. (2005). Atorvastatin (therapy for hyperlipidemia) may dose-dependently impair erectile function. (submitted)

[111] Woo, JT; Nakagawa, H; Krecic, AM; Nagai, K; Hamilton, AD; Sebti, SM; Stern, PH. (2005). Inhibitory effects of mevastatin and a geranylgeranyl transferase I inhibitor (GGTI-2166) on mononuclear osteoclast formation induced by receptor activator of NF kappa B ligand (RANKL) or tumor necrosis factor-alpha (TNF-alpha). *Biochem Pharmacol, 69,* 87-95.

[112] Wu, J; Wong, WW; Khosravi, F; Minden, MD; Penn, LZ. (2004). Blocking the Raf/MEK/ ERK pathway sensitises acute myelogenous leukaemia cells to lovastatin-induced apoptosis. *Cancer Res, 64,* 6461-6468.

[113] Yaffe, K; Barrett-Connor, E; Lin, F; Grady, D. (2002). Serum lipoprotein levels, statin use, and cognitive function in older women. *Arch Neurol, 59,* 378-384.

[114] Yang, CC; Jick, SS; Jick, H. (2003). Lipid-lowering drugs and the risk of depression and suicidal behaviour. *Arch Intern Med, 8,* 1926-1932.

[115] Young-Xu, Y; Chan, KA; Liao, JK; Ravid, S; Blatt, CM. (2003). Long-term statin use and psychological well-being. *J Am Coll Cardiol, 20,* 690-697.

In: Statins: Indications and Uses, Safety and Modes of Action ISBN 978-1-60692-103-6
Editor: Gunnar N. Holmquist

Chapter 3

PRELIMINARY FINDINGS ABOUT THE TRENDS OF THE RENAL HEMODYNAMICS AND THE PROXIES OF CARDIOVASCULAR RISK DURING A SHORT COURSE OF ATORVASTATIN THERAPY IN ESSENTIAL HYPERTENSIVES

Luigi Vernaglione
Nephrology and Dialysis Unit 'M. Giannuzzi'Hospital- Manduria, Italy

ABSTRACT

The statins are widely used for the prevention of cardiovascular diseases thanks to their lipid lowering effects. Recent reports suggest that statins have antihypertensive and antiinflammatory properties besides the lipid lowering one. Moreover, in animal models of hypertension and kidney disease, the statins induced both the eNOS up-regulation and the iNOS, LOX-1 and NFkB down-regulations in the renal endothelium promoting the increase of the perfusion lowering the sclerosis as well. All these effects of statins might be helpful in reducing the incidence of cardiovascular diseases in human hypertensives whose clinical picture is often characterized by risk factor such as dyslipidemia, elevated serum C-reactive protein levels and nephrosclerosis with reduced renal vascular reserve and microalbuminuria.

The aim of this prospective, self-controlled, interventional study was to assess the trends of the renal hemodynamics, the blood pressure, the lipid panel and the inflammation during a short course of atorvastatin therapy, in essential hypertensives (EH).

After a 5-days run-in period, the patients began atorvastatin therapy (10 mg q.d.). At T_0 (before the first tablet of atorvastatin) and T_{30} (after 30 days of therapy) the patients underwent the following assessments: systolic, diastolic and mean blood pressure (SBP, DBP and MAP respectively), glomerular filtration rate (GFR) and effective renal plasma flow (ERPF) (inulin and p-aminohippurate clearance respectively), filtration fraction

(FF) (GFR/ERPF) and total renal vascular resistances (TRVR) (using ERPF, hematocrit and MBP), microalbuminuria and serum levels of lipids and C-reactive protein (hs-CRP). The patients followed a standard diet (35 Kcal/kg bw/day, 1 g of protein/kg bw/day and 6 g of sodium/day). Never treated patients with I stage EH were included. Patients with heart, kidney or liver diseases were excluded as well as patients taking any other medication.

This report provides preliminary data about 6 EH (3M/ 3F), 45.5 ± 14.8 (mean ± SD) years old so far enrolled. The patients have been compliant with the atorvastatin therapy as showed by the reduction of serum lipids (mg/dL) (total cholesterol: T_0: 226.8 ± 33.7 vs T_{30}: 164.5 ± 46.5, p=0.01; triglycerides: T_0: 157.8 ± 50.2 vs T_{30}: 89.8 ± 16.5, p=0.05; apo-B: T_0: 110.3 ± 24.2 vs T_{30}: 75.5 ± 28.8, p=0.003). Mean daily protein and sodium intake as well as serum hs-CRP levels, microalbuminuria and renal hemodynamics were unchanged during the study instead of blood pressure which was significantly reduced (mmHg) (SBP: T_0 : 150.3 ± 10.6 vs T_{30}: 139.1 ± 6.9, p=0.01; DBP: T_0 : 97.4 ± 11.5 vs T_{30}: 88.6 ± 9.4, p=0.04; MAP: T_0 : 115.0 ± 11.0 vs T_{30}: 105.5 ± 7.9, p=0.02).

This preliminary experience on EH suggest that in EH a short course of atorvastatin therapy is associated with the statistically significant reduction of both the serum lipids levels and the blood pressure and the restoration of the glomerular autoregulation. The absence of changes of the serum hs-CRP levels and microalbuminuria during atorvastatin therapy might mean that the pleiotropic effects of statins demonstrated in animal models might be subdivided in short-term and long-term effects, depending on the type, the dose and the timing of therapy.

INTRODUCTION

It has been shown that the treatment of hypercholesterolemia with 3-hydroxy-3-methylglutaryl CoA reductase inhibitors (statins) reduces significantly the incidence of cardiovascular events [1,2]. Moreover high serum cholesterol levels are nowadays considered as independent risk factor for renal diseases [3,4].

Recent reports suggest that the statins have antihypertensive and antiinflammatory properties besides the lipid lowering one. For example in animal models of hypertension and kidney disease, the statins had been able to induced both the eNOS up-regulation and the iNOS, LOX-1 and NFkB down-regulations in the renal endothelium. These changes allowed for the increase of nitric oxide availability which induced the glomerular vasodilation. Moreover the nitric oxide contrasts the oxidative stress and the fibroblasts proliferation at the same time [5,6].

In vitro and in vivo studies have demonstrated that the statins reduce the cytokines synthesis and the mesangial proliferation in the kidney [7]. Moreover the statins have been shown able to improve the kidney function and to reduce the blood pressure and the serum cholesterol levels in experimental models of nephropathy [7-11].

In a recent experience on the pigs the administration of simvastatin rose the eNOS expression in the renal endothelial cells and did improve the renal perfusion, without influencing the serum cholesterol levels [12]. Those hemodynamical effects have been explained by the increase of nitric oxide synthesis and the neutralization of the superoxide anions in renal vasculature.

Even the atorvastatin showed the property to influence both the nitric oxide regulation and the cytokine signaling in humans and animals [13-16].

One of the most frequent causes of chronic kidney disease in the western countries is the hypertensive nephroangiosclerosis. According to the most accepted theory both the shear-stress and the stretch stress on the vessel walls induced by the changes in the systemic and glomerular hemodynamics play a role in the determinism of that nephropathy by promoting the overproduction of cytokines which quickens the glomerulosclerosis worsening the glomerular ischaemia as well [17].

If the impact of statins on the cytokine signaling is confirmed in well powered clinical studies it will be possibile to impact the pathways described above slowing down the progression of the nephroangiosclerosis and improving the renal perfusion.

All the lipid-independent effects of statins depicted above might be helpful in reducing the incidence of the cardiovascular diseases in human hypertensives whose clinical picture is often characterized by risk factor such as dyslipidemia, high serum C-reactive protein levels and nephrosclerosis with reduced renal vascular reserve and microalbuminuria.

The main aim of this study was to verify if, in EH, a short course of atorvastatin therapy is associated with changes of the renal hemodynamics as well as of the trends of proxies of cardiovascular morbidity such as hypertension, high serum CRP and lipids levels and microalbuminuria. Moreover the safety of the atorvastatin therapy in terms of liver and muscle metabolisms was investigated.

SUBJECTS AND METHODS

a) Patients

I stage EH never treated patients (according to the JNC VII) of both genders, with normal renal and liver functions and without dyslipidemia, proteinuria and other comorbidities were considered eligibles for the study.

Subjects suffering from kidney and heart diseases or secondary hypertension as well as those who had experienced liver, muscle or inflammatory diseases, malignancies, surgery, traumatisms, infections and therapy with steroids within the last 3 months from the beginning of the study were excluded.

Moreover we foresaw the exclusion from the study of those subjects presenting the above mentioned conditions during the follow-up.

b) Study Design

We designed a prospective, 35 days, self controlled, interventional study during which each subject underwent a 5-days screening period, for the evaluation of the inclusion/exclusion criteria, followed by a double checking of renal hemodynamics, blood pressure and biochemistry parameters:

at T_0: first day of study (just the day after the end of the screening period, before taking the study drug);

at T_{30}: after 30 days of oral atorvastatin administration (10 mg q.d.);

We chose to test the atorvastatin because it already showed vasodilating endothelium-related properties both on animals and in humans [13-16]. Moreover the atorvastatin pharmacokinetics warranted a better 24-hours coverage with lower amounts of drug [16].

Starting from the month before the study the subjects enrolled followed a standardized diet with 35 Kcal/kg b.w/day, daily protein intake of 1.0 g/kg b.w. and daily sodium intake of 6 g, in order to control all the main nutritional influences on the renal hemodynamics. Moreover during the study the patients did not take any medications.

The compliance of the subjects to the atorvastatin was double checked by means of the subjective daily reports and the serum lipid levels monitoring.

c) Parameters Evaluated

After giving the informed consent, at T_0 and T_{30} the subjects underwent the following evaluations:

1. continous monitoring of systolic, diastolic and mean blood pressure (SBP, DBP and MBP) (ABPM Monitor 90207, Spacelabs Medical Inc, Redmond, WA);
2. glomerular filtration rate (GFR) and effective renal plasma flow (ERPF) by means of the inulin and p-aminohippurate clearances respectively (see below);
3. filtration fraction (FF) and total renal vascular resistances (TRVR) (see below);
4. body mass index (BMI) (with the formula body weight (kg)/height^2 (m));
5. serum levels of albumin (nephelometric method, Dade Behring-Marburg GmbH, Germany), total cholesterol and triglycerides (enzymatic determination, ABX Diagnostics, Montpellier, France), apo-A and B (immunonephelometric method, Dade Behring-Marburg GmbH, Germany), alanine transferase (ALT), aspartate transferase (AST), gamma glutamyltransferase (GGT), creatinphosphokinase (CPK), lactate dehydrogenase (LDH) (kinetic determination, ABX Diagnostics, Montpellier, France) and hs-CRP (N high sensitivity immunonephelometric method, Dade Behring-Marburg GmbH, Germany). All the listed dosages, except for CRP, albumin and hematocrit ones, were performed utilizing a Cobas Mira Plus (Roche, Italy) automatic instrument. The methodology used for the assessment of serum CRP levels is based on the agglutination of polystyrene particles coated with monoclonal antibodies to CRP when mixed with samples containing CRP. This method is designed to measure CRP concentrations within an overall range of 0.175 to 1100 mg/L, with a normal value < 3 mg/L. The typical limit of detection for CRP is 0.175 mg/L for measurement performed using a sample dilution of 1:20, with a coefficient of variation $<4.4\%$ for intra-assay precision and $<5.7\%$ for inter-assay reproducibility. The regression analysis of this method (y) with the N Latex CRP kit or ELISA Hemagen method (x) yielded respectively the following equations: $y = 0.95x + 1.43$ (mg/L) and $y = 0.75x - 0.25$ (mg/L);

6. urine microalbumin levels (turbidimetric method, Dade Behring-Marburg GmbH, Germany);
7. 24 hours urine nitrogen (enzymatic determination, ABX Diagnostics, Montpellier, France) and sodium levels (direct potentiometry, ABX Diagnostics, Montpellier, France) in order to calculate the daily sodium intake, with the formula: 0.06 x urine sodium mEq/L, and the daily protein intake with the formula: 6.25 x (24 h urine nitrogen mg% + 30mg/kg).

The values considered for the statistical analysis derived from the average of 2 consecutive evaluations performed within 48 hours.

d) Assessment of the renal hemodynamical parameters

The clearance assay was carried out in the morning in the fasting state. At –30 min. the patients underwent bladder catheterization and, after bladder emptying, a mineral water oral intake of 15 ml/kg body weight in order to optimize the hydratation status. Moreover, at the same time a continue blood pressure monitoring was started. At –15 min. a i.v. bolus of inulin and p-aminohippurate (provided by Jacopo Monico, Mestre, Italy) was started. This infusion lasted 15 minutes. At time 0 min.a continuous i.v. drip of inulin and p-aminohippurate was begun (Pump Model 4, Abbot Laboratories, North Chicago IL 60064 USA). Either the bolus and the drip were made according to the Duarte's formulas [18]. After a period of 60 min., during which the patient underwent an oral water replacement according to the urine output, three consecutive clearances were tested, each calculated over 30 (from time 60 min to time 150 min). During the clearances period the blood for hematocrit (Coulter Counter Gen*S1, Instrumentation Laboratory, Italy) and the serum inulin and p-aminohippurate levels assessments was drawn at 75, 105 and 135 minutes. Moreover an urine sampling for the assessment of the inulin and p-aminohippurate levels was performed at 90, 120 and 150 minutes. At the same times the urine output was recorded. The session finished at 150 min.

The day of the hemodynamical study an antibacterial prophylaxis was given consisting of ciprofloxacin 500 mg/day to avert the risk of urinary infection from the bladder catheterization.

Inulin and p-aminohippurate were assayed with the colorimetric method using diphenylamine [19] and ethylenediamine-N-1-naphthyl chlorhydrate [20] respectively (see appendix).

Thus the renal hemodynamical parameters were calculated by using the following formulas:

- GFR: [urine inulin level (mg/dL) x urine output (ml/min)]/ serum inulin level (mg/dL);
- ERPF: [urine p-aminohippurate level (mg/dL) x urine output (ml/min)]/ serum p-aminohippurate level (mg/dL);
- FF: GFR/ERPF

- TRVR: [MAP/effective renal blood flow (ERBF)], where ERBF= ERPF/[1-(hematocrit/100)].

The values of the renal hemodynamical parameters used in the present analysis derived from the average of the 3 consecutive clearances of inulin and p-aminohippurate measured in each session.

e) Statistical Analysis

The distribution of the data was studied by means of the Kolmogorov-Smirnov test. Then the Student's t-test for paired data was performed in order to compare the average values of the normally distributed parameters at T_0 and T_{30} The same comparison was performed using the Wilcoxon's test for the non-normally distributed parameters. Values of $p < 0.05$ were considered as statistically significant. The statistical inference was performed by means of the SPSS software 10.1 (SPSS Inc. Chicago, IL).

RESULTS

In this paper we report a set of preliminary data regarding 6 EH, (3M and 3F), 45.5 ± 14.8 years old, with MAP of 115.0 ± 11.0 mmHg, so far enrolled.

The subjective daily reports provided by these subjects showed regular administrations of the atorvastatin therapy. Moreover, the statistically significant reductions of the serum total cholesterol (p=0.01), triglycerides (p=0.05) and apo-B (p=0.003) levels between T_0 and T_{30} presented in table 1 and expected during atorvastatin therapy, suggest a satisfactory compliance of the subjects to the study drug.

Table 1. Comparison between the nutritional parameters (average ± SD) at T_0 and T_{30}

	T_0	T_{30}
s. triglycerides (mg/dL)	157.8 ± 50.2	89.8 ± 16.5 [a]
s. total cholesterol (mg/dL)	226.8 ± 33.7	164.5 ± 46.5 [b]
s. apo A (mg/dL)	153.8 ± 20.5	153.0 ± 9.8
s. apo B (mg/dL)	110.3 ± 24.2	75.5 ± 28.8 [c]
BMI (kg/m^2)	23.9 ± 4.2	22.6 ± 3.7
s. albumin (g/dL)	4.3 ± 0.2	4.5 ± 0.2
sodium intake (g/day)	6.8 ± 1.4	7.6 ± 1.7
protein intake (g/day)	1.4 ± 0.1	1.3 ± 0.3

[a] Student's t-test for paired data: p = 0.05;
[b] Student's t-test for paired data: p = 0.01;
[c] Student's t-test for paired data: p = 0.003.

The table 1 also shows the average values of BMI, serum albumin levels, daily sodium and nitrogen intake of the subjects at T_0 and T_{30}. The comparison of these data do not reveal any statistically significant difference meaning that the subjects held a costant nutritional schedule according to the standardized diet. Thus this figure suggests that in the study the food habits of the subjects did not influence the outcomes of interest.

In table 2 the average values of GFR, ERPF, FF and TRVR presented by the subjects at T_0 and T_{30} are depicted. The analysis of these data demonstrates that the renal hemodynamics was stable during the follow-up (p>0.05 for each comparison).

During the follow-up the subjects studied showed up also unchanged levels of serum hs-CRP and microalbuminuria (mg/L; average $\pm$ SD): 19.6 $\pm$ 15.4 (T_0) vs 21.4 $\pm$ 13.3 (T_{30}) (p=0.90) and (mg/L; median and range) 6 (6 - 21) vs 6 (3 – 24); p=0.50.

Table 2. Comparison between the renal hemodynamical parameters (average $\pm$ SD) at T_0 and T_{30}

	T_0	T_{30}
GFR (mL/min/1,73m^2)	57.5 $\pm$ 15.4	65.8 $\pm$ 17.3
ERBF (mL/min/1,73 m^2)	357.1 $\pm$ 74.3	364.6 $\pm$ 58.1
FF	0.16 $\pm$ 0.05	0.18 $\pm$ 0.03
TRVR (dynes x sec^{-1} x cm^{-5})	13992 $\pm$ 4680	12965 $\pm$ 3191

Student's t-test for paired data: p > 0.05 for each comparison.

Table 3. Comparison between the blood pressure values (average $\pm$ SD) at T_0 and T_{30}

	T_0	T_{30}
SBP (mmHg)	150.3 $\pm$ 10.6	139.1 $\pm$ 6.9^a
DBP (mmHg)	97.4 $\pm$ 11.5	88.6 $\pm$ 9.4^b
MAP (mmHg)	115.0 $\pm$ 11.0	105.5 $\pm$ 7.9^c

[a] Student's t-test for paired data: p = 0.01
[b] Student's t-test for paired data: p = 0.04
[c] Student's t-test for paired data: p = 0.02

Table 4. Comparison between the liver and muscle enzymes (median and range) at T_0 and T_{30}

	T_0	T_{30}
ALT (U/L)	16 (15 – 26)	13,5 (13 – 16)
AST (U/L)	13,5 (10 – 17)	12 (7 – 20)
GGT (U/L)	10 (7 – 24)	11 (5 – 37)
LDH (U/L)	154 (130 – 173)	159,5 (144 – 165)
CPK (U/L)	110 (55 – 126)	128 (77 – 246)

Wilcoxon's test: p>0.05 for each comparison.

As reported in table 3 the 30-days course of atorvastatin was associated with statistically significant reductions of SBP (p=0.01), DBP (p=0.04) and MAP (p=0.02).

Finally the table 4 shows the average values of the liver and muscle enzymes of the subjects at T_0 and T_{30}. The comparison between these data do not reveal any significant changes (p>0.05 for each comparison).

Discussion

Recent reports suggest that statins have many other pharmacological properties besides the lipids lowering one. In the next future such a properties could widen the employment of statins in the prevention of the cardiovascular diseases. Actually several studies have demonstrated that statins slow down the atherosclerosis by interfering with the endothelial cells apoptosis, the fibroblasts action, the nitric oxide pathway, the cytokines network and, finally, the lipid metabolism [5,6,21].

Studies performed both in the animals and in humans have shown that simvastatin and atorvastatin are able to stimulate the endothelial synthesis of nitric oxide inhibiting that of inflammation and fibrosis-related cytokines as well [12-16]. However, to date, the literature lacks investigations focused on the actions of the statins on both the systemic and the renal hemodynamics. If their effects on the nitric oxide and cytokines pathways are able to influence the plasticity of the vessel walls, the statins will induce changes in the hemodynamics that could play an important role in the treatment of hypertension and its related organ damage.

In this study, focused on the EH, we have assessed if a short course of atorvastatin therapy is associated with changes of the renal hemodynamics. Moreover we have analyzed the trends of some proxies of cardiovascular risk such as hypertension, high serum CRP and lipids levels and microalbuminuria during the atorvastatin treatment. Finally we have investigated the safety of the atorvastatin therapy.

Although this report presents only preliminary findings, our analysis provides some important suggestions.

First of all, in our experience, the administration of atorvastatin has been well tolerated. Actually, all the subjects so far tested in the study were compliant to the therapy as demonstrated by the subjective daily reports as well as the statistically significant reduction of the serum lipids levels. Moreover the subjects did not suffer from side effects potentially attributable to the atorvastatin therapy. In this context it is also important to underline that the improvement of the lipidic profile obtained in our population during the study should not be attributable to changes of food habits, provided that BMI, serum albumin levels and both sodium and protein intakes resulted similar at T_0 and T_{30}.

The main result of our analysis has been that the EH studied presented a statistically significant reduction of all the components of the blood pressure monitoring (SBP, DBP and MAP), coupled with the absence of any changes of the renal hemodynamics, during the atorvastatin administration period (10 mg q.d. for 30 days).

To date the knowledges about the relationships between statins and renal hemodynamics are quite scarce. However our result in humans contrasts with those of the two reports focused on this topic, in which an increase of the renal blood flow in the course of lovastatin therapy is described in rodents [22-23]. Even if the choice of the dose and the timing of the statin therapy as well as the study subjects could represent the causes of these contrasts, in our opinion the main reason of the discrepancy resides in the fact that the evaluation of the renal hemodynamics cannot be separated from the analysis of the blood pressure trend and its relationship with the glomerular arterioles tone autoregulation.

In normal subjects the GFR and the ERPF are stable in front of changes of the blood pressure, in the range 80-180 mmHg. This because the tunica media of both the afferent and efferent arterioles autoregulates their vascular tone and so by doing keeps constant the intraglomerular pressure and the blood flow. The renin-angiotensin system, the prostanoids pathway and the adenosine metabolism might be also involved in this mechanism [24]. In essential hypertensives the loss of the autoregulation allows for the transmission of the high blood pressure levels to the glomerular tuft. This transmission drives the rise of both the GFR and the ERPF coupled with the increase of the shear and the stretch stress on the glomerular walls. This chain of events might be the basis for the development of the glomerular sclerosis in EH [25]. Disorders of the endothelial nitric oxide pathway have been proposed as the cause of the loss of the glomerular autoregulation in EH [26].

In our population of EH, during the atorvastatin therapy, the renal hemodynamics did not change in front of a significant fall of the blood pressure. This means that during the follow-up our hypertensive subjects presented the glomerular autoregulation. Taking into account the endothelial effects of statins on the nitric oxide pathway demonstrated in the experimental models we could speculate that in our experience the atorvastatin might have restored the glomerular autoregulation of the subjects by influencing the endothelial nitric oxide expression. This hypotesis will be tested in the next steps of the present study in which we are going to check for the blood nitric oxide levels and other endothelial markers.

The clinical picture of the EH is often characterized by risk factors for cardiovascular diseases other than hypertension. The pleiotropic effects of the statins demonstrated in experimental models open fascinating perspectives about the usefulness of these medication, for aims other than the lipidic control, in the primary or secondary prevention of the cardiovascular diseases.

In this study we focused the attention on some cardiovascular risk factors, finding that the atorvastatin therapy is associated with the reductions of the lipids levels and the blood pressure. Neither the serum CRP levels nor the microalbuminuria changed during the administration of the study drug. These figures suggest that the pleiotropic (antiinflammatory, antifibroblast, anti oxidative stress, etc) effects of the atorvastatin (and other statins) might be subdivided in short-term and long-term effects, also depending on the type, the dose and the timing of therapy. Thus studies focused on the confirmation of the pleiotropic effects of the statins in humans should be randomized, placebo-controlled and performed by using escalating doses of the medication and longer follow-up.

Finally, the liver and muscle toxicity are the well known side effects of the statins. In our experience the short course of atorvastatin therapy has been safe and well tolerated by the subjects. This statement is based upon the fact that both the liver and the muscle enzymes tested during the study were unmodified.

In conclusion our preliminary results suggest that in EH a short course of atorvastatin therapy is associated with the statistically significant reduction of both the serum lipids levels and the blood pressure and the restoration of the glomerular autoregulation. Nonetheless, the absence of effects of the atorvastatin on the serum CRP levels and microalbuminuria undelines that the pleiotropic effects of the statins in humans might be subdivided in short-term and long-term effects, also depending on the type, the dose and the timing of therapy Thus other placebo-controlled, randomized studies performed with higher number of patients, larger follow-up and escalating doses of medication are warranted.

APPENDIX

Determination of the Inulin:

a) Serum

- After centrifugation of the blood (3000 RPM) 300 μL of serum were diluted with 4.5 mL of distilled water;
- 600 μl of a water solution of sodium hydroxide (0.75N) and 600 μl of a water solution of zinc sulphate (10%) were added;
- the specimen was centrifuged for 20 minutes (3000 RPM)
- 250 μL of a water solution of sodium hydroxide (4N) were added to 1 mL of surnatant;
- after 10 minutes 4 mL of diphenylamine solution (14 mg/mL in glacial acetic acid and hydrochloric acid) were added;
- after 1 hour the specimen was assayed by the spectrophotometer (wavelenght 620 nm).

b) Urine

- 100 μL of urine were diluted with distilled water according to the urine output obtained during the session;
- 250 μL of a water solution of sodium hydroxide (4N) were added to 1 mL of specimen;
- after 10 minutes 4 mL of diphenylamine solution (14 mg/mL in glacial acetic acid and hydrochloric acid) were added;
- after 1 hour the specimen was assayed by the spectrophotometer (wavelenght 620 nm).

Determination of the p-Aminohippurate:

a) Serum

- After centrifugation of the blood (3000 RPM) 300 μL of serum were diluted with 3 mL of distilled water;
- 6 mL of trichloroacetic acid solution (10%) were added;
- the specimen was centrifuged for 20 minutes (3000 RPM)

- 200 μL of a water solution of sodium nitrite (0.1%) were added to 1 mL of surnatant;
- 200 μL of a water solution of ammonium sulphamate (0.5%) were added to the specimen;
- 600 μL of a water solution of ethylenediamine-N-1-naphthyl chlorhydrate (0.1%) were added to the specimen;
- after 15 minutes the specimen was assayed by the spectrophotometer (wavelenght 545 nm).

b) Urine

- 100 μL of urine were diluted with distilled water according to the urine output obtained during the session;
- 300 μL of specimen were diluted with 3 mL of distilled water;
- 6 mL of trichloroacetic acid solution (10%) were added;
- the specimen was centrifuged for 20 minutes (3000 RPM)
- 200 μL of a water solution of sodium nitrite (0.1%) were added to 1 mL of surnatant;
- 200 μL of a water solution of ammonium sulphamate (0.5%) were added to the specimen;
- 600 μL of a water solution of ethylenediamine-N-1-naphthyl chlorhydrate (0.1%) were added to the specimen;
- after 15 minutes the specimen was assayed by the spectrophotometer (wavelenght 545 nm).

REFERENCES

[1] Bonetti PO, Lerman LO, Napoli C, Lerman A: Statins effects beyond lipid lowering--are they clinically relevant?. *Eur Heart J 24*(3): 225-248, 2003

[2] MRC/BHF Heart Protection Study of cholesterol lowering with simvastatin in 20 536 high-risk individuals: a randomized placebo-controlled trial. *Lancet 60*: 7-22, 2002

[3] Vazquez-Perez S, Aragoncillo P, de Las Heras N, Navarro-Cid J, Cediel E, Sanz-Rosa D, Ruilope LM, Diaz C, Hernandez G, Lahera V, Cachofeiro V: Atorvastatin prevents glomerulosclerosis and renal endothelial dysfunction in hypercholesterolaemic rabbits. *Nephrol Dial Transplant 16* (1): 40-44, 2001

[4] Fried LF, Orchard TJ, Kasiske BL: Effect of lipid reduction on the progression of renal disease, a meta-analysis. *Kidney Int 59*: 260-269, 2001

[5] Maron DJ, Fazio S, Linton MF: Current perspectives on statins. *Circulation 101*: 207-213, 2000

[6] Wilson SH, Simari RD, Best PJ, Peterson TE, Lerman LO, Aviram M, Nath KA, Holmes DR Jr, Lerman A: Simvastatin improves coronary endothelial function independent of lipid lowering. *Arterioscler Thromb Vasc Biol 21*: 122-128, 2001

[7] Buemi M, Senatore M, Corica F, Aloisi C, Romeo A, Cavallaro E, Floccari F, Tramontana D, Frisina N: Statins and progressive renal disease. *Med Res Rev 22*: 76-84, 2002

[8] O'Donnel MP, Kasiske BL, Kim Y, Schmitz PG, Keane WF: Lovastatin retards the progression of established glomerular disease in obese Zucker rats. *Am J Kidney Dis 22*: 83-89, 1993

[9] Wilson TW, Alonso-Galicia M, Roman RJ: Effects of lipid-lowering agents in the Dahl salt-sensitive rat. *Hypertension 31*: 225-231, 1998

[10] Hafez KS, Inman SR, Stowe NT, Novick AC: Renal hemodynamic effects of lovastatin in a renal ablation model. *Urology 48*: 862-867, 1996

[11] Park JK, Muller DN, Mervaala EM, Dechend R, Fiebeler A, Schmidt F, Bieringer M, Schafer O, Lindschau C, Schneider W, Ganten D, Luft FC, Haller H: Cerivastatin prevents angiotensin II-induced renal ingiury independent of blood pressure and cholesterol-lowering effects. *Kidney Int 58*: 1420-1430, 2000

[12] Wilson SH, Chade AR, Feldstein A, Sawamura T, Napoli C, Lerman A, Lerman LO: Lipid-lowering-independent effects of simvastatin on the kidney in experimental hypercholesterolaemia. *Nephrol Dial Transplant 18*: 703-709, 2003

[13] Asberg A, AHartmann A, Fjeldsa E, Holdaas H: Atorvastatin improves endothelial function in renal-transplant recipients. *Nephrol Dial Transplant 16*: 1920-1924, 2001

[14] Pisani A, Uccello F, Cesaro A, Comi N, Mirenghi F, Serio V, Sabbatini M: Effetti dell'atorvastatina nell'insufficienza renale acuta ischemica del ratto anziano. *Giorn It Nefrol 19* (5): 534-539, 2002

[15] Tan KC, Chow WS, Tam SC, Ai VH, Lam CH, Lam KS: Atorvastatin lowers C-reactive protein and improves endothelium-dependent vasodilation in type 2 diabetes mellitus. *J Clin Endocrinol Metab 87*(2): 563-568, 2002

[16] Malinowski JM: Atorvastatin: a hydroxymethylglutaryl coenzyme A reductase inhibitor. *Am J Health Syst Pharm 55*(21): 2253-2267, 1998

[17] Dworkin LD, Hostetter TH, Renke HG, Brenner BM: Hemodynamic basis for glomerular injury in rats with desoxycorticosterone-salt hypertension. *J Clin Invest 73*: 1448, 1984

[18] Liedtke RR, Duarte CG: Laboratory protocols and methods for the measurement of glomerular filtration rate and renal plasma flow. In *Renal function test: clinical laboratory procedures and diagnosis,* edited by Duarte CG, Boston, Little Brown Company, 1980, pp 49-63

[19] Walser M, Davidson DG, Orloff J: The renal clearance of alkali-stable inulin. *J Clin Invest 34*: 1520-1523, 1955

[20] Hamburger J, Richet G, Crosnier J, Funck-Brentano JL, Antoine B, Ducrot H, Méry JP, De Montera H: Méthodes chimiques et physiques utilisées en exploration fonctionnelle. In *Néphrologie,* edited by Hamburger J, Richet G, Crosnier J, Funck-Brentano JL, Antoine B, Ducrot H, Méry JP, De Montera H, Paris, Flammarion, 1966, pp 274-276

[21] Buemi M, Senatore M, Corica F, Aloisi L, Romeo A, Cavallaro E, Frisina N: Are the potential non-lipid lowering uses of statins in the kidney? *Nephron 89*: 363-368, 2001

[22] Hafez KS, Inman SR, Stowe NT, Novick AC: Renal hemodynamic effects of lovastatin in a renal ablation model. *Urology 48*(6):862-867, 1996

[23] Stowe NT, Inman SR, Tapolyai M, Brouhard BH, Hodge EE, Novick AC: Lovastatin has direct renal hemodynamic effects in a rodent model. *J Urol. 156*(1):249-252, 1996

[24] Vander AJ: Regulation of the renal circulation. In *Renal Physiology,* edited by Vander AJ, Milan, McGraw-Hill, 1991, pp 61-74

[25] Beaufils M: Preserving the autoregulation of renal hemodynamics. *Cardiology 83* (Suppl 1):10-15, 1993

[26] Klahr S, Morrissey J: L-arginine as a therapeutic tool in kidney disease. *Semin Nephrol. 24*(4):389-394, 2004

In: Statins: Indications and Uses, Safety and Modes of Action ISBN 978-1-60692-103-6
Editor: Gunnar N. Holmquist

Chapter 4

BENEFICIAL EFFECTS OF THE ADDITION OF FENOFIBRATE TO STATIN THERAPY IN PATIENTS WITH ACUTE CORONARY SYNDROME AFTER PERCUTANEOUS CORONARY INTERVENTIONS

Hetal D. Shah[1], Keyur H. Parikh[2], Milan C. Chag[2], Urmil G. Shah[2], Hemang A. Baxi[2], Anish H. Chandarana[2], Ajay M. Naik[2], Joyal N. Shah[2], Kanan J. Shah[1] and Ramesh K Goyal[1*]

[1]Department of Pharmacology, L. M. College of Pharmacy Navrangpura, Ahmedabad 380 006 Gujarat, India

[2]The Heart Care Clinic, Care Cardiology Consultants Pvt Ltd 201, Ahmedabad 380 009, Gujarat, India

ABSTRACT

Objective: The objective of the present investigation was to find out whether the addition of fenofibrate to statin monotherapy produced any synergistic or additive beneficial effects in reducing risk factors, especially plasma fibrinogen, in patients of Acute Coronary Syndromes (ACS) requiring Percutaneous Coronary Interventions (PCI).

Methods: This was a randomized, non-blind, prospective study with parallel group design, conducted in 102 patients who had angiographically documented Coronary Artery Disease (CAD). All had undergone angioplasty. The patients were randomized to atorvastatin (20mg/day, n=25), simvastatin (40mg/day, n=27), atorvastatin(10mg/day)-fenofibrate (200mg/day) combination (n=25) and simvastatin (20mg/day)-

* Address for Correspondence and Reprints: Professor Ramesh K Goyal, Department of Pharmacology, L. M. College of Pharmacy Navrangpura, Ahmedabad 380 006 Gujarat, INDIA. Phone: 91-79-26302746. Email: goyalrk@hotmail.com, hetald_shah@yahoo.com.

fenofibrate(200mg/day) combination (n=25). The serum lipid profile and plasma fibrinogen were recorded before initiation of therapy and after 3 months of the respective treatments.

Results: All the patients already had desirable lipid levels as per the NCEP ATP III guidelines. The addition of fenofibrate to statin monotherapy produced additional benefits on reduction in triglyceride (TG) and very low density lipoprotein (VLDL) levels, and an increase in high density lipoprotein (HDL) levels. All the treatment groups showed a significant decrease in the plasma fibrinogen levels. This did not correlate with any of the study parameters like age, body weight, hemodynamic characteristics and lipoprotein levels. Statin monotherapy produced a significant decrease in the fibrinogen levels and the addition of fenofibrate further enhanced the reduction.

Conclusions: Addition of fenofibrate to statins seems to be beneficial in patients with ACS. Statins, contrary to various reports, were found to decrease plasma fibrinogen significantly.Furhter, in combination with fenofibrate there was enhanced reduction of the novel risk factor, fibrinogen.

Keywords: Acute coronary syndromes, statins, fenofibrate, plasma fibrinogen

INTRODUCTION

The last century has seen a rapid increase in the global prevalence of Coronary artery disease (CAD). Estimates from the Global Burden of Disease Study have predicted that India faces the greatest health burden due to Coronary Artery Disease.[1] A number of inflammatory markers have been studied for their ability to predict future cardiovascular events in asymptomatic individuals and patients with established atherosclerotic disease.[2] Among the emerging novel cardiac markers, plasma fibrinogen has been identified as an important risk factor for cardiovascular diseases. Many cross-sectional, case controlled, and numerous prospective cohort studies have identified elevated plasma fibrinogen levels as an independent risk factor for coronary heart disease, stroke, and peripheral vascular disease [3]

Fibrinogen is an acute phase protein that is directly involved in coagulation. The transcription of fibrinogen is stimulated by IL-6; its synthesis is suppressed by IL-1β and TNF-α. [4,5] Fibrinogen and its metabolites strongly affect hemostasis, hemorheology, platelet aggregation and endothelial function. Infact, fibrinogen's association with increased mortality lies probably in its ability to promote thromboses, or clots, by causing platelet aggregation in blood vessels. The recognition that fibrinogen is an important factor in the promotion of various disease states has led to the search for specific therapies intended to reduce plasma fibrinogen levels.

The use of statins in the prevention of primary [6] and secondary [7,8] CAD has been demonstrated to significantly reduce cardiovascular events and total mortality. Nevertheless, the majority of patients on statin treatment still experience coronary events. [9] It is clear that a more effective reduction in the incidences of coronary events is needed. This could probably be accomplished by a further reduction in the conventional as well as novel risk factors like plasma fibrinogen, homocysteine, and C-reactive protein. Multiple small studies have reported changes in fibrinogen levels with different statins. [10, 11] While atorvastatin is claimed to increase plasma fibrinogen, simvastatin is reported to either increase or have a neutral effect. Although many different pharmacologic approaches and strategies for therapeutic modulation of fibrinogen have been tested, the efficacy of different treatments to lower plasma fibrinogen in humans is limited and the mode of action unidentified. [12, 13] Among the few compounds known to lower circulating fibrinogen levels in humans are certain fibrates.

Fibrates reportedly lower plasma fibrinogen in humans, but the regulatory mechanism of this effect remains to be clarified. Activation of the nuclear hormone receptor PPARα mediates the suppression of fibrinogen gene transcription by fibrates in rodents.[3] This establishes PPARα as a key regulatory factor in fibrinogen gene expression in rodents and may explain the suppressive effect of fibrates on plasma fibrinogen levels in humans. The fibrinogen molecule is arranged as a dimer, with each monomer composed of 3 nonidentical polypeptide chains: Aα, Bβ and γ. [14] The 3 fibrinogen chains are encoded by 3 separate, closely linked genes situated on the same chromosome and located in the sequence γ, Aα, and Bβ, with the last one in opposite transcriptional orientation to the first one.[15] Binsack et al [16] reported that in the human hepatoma cell line, HepG2, bezafibrate suppressed Aα-, Bβ- and γ- chain mRNA levels.[17] The genes encoding the 3 fibrinogen chains are negatively regulated by PPARα and since fibrates act through this receptors, it helps to explain the benefits of fibrate therapy to target a reduction in fibrinogen levels. Maison et al[18] showed that fibrinogen concentration decreased after fibrate therapy, while it increased after statin treatment. Thus this study was conducted to investigate the controversial role of statins in modifying fibrinogen levels and to study the benefits of addition of fenofibrate to statin therapy in modifying levels of plasma fibrinogen.

Research Design and Methods

The study was a controlled clinical trial. It was randomized, non blind and based on parallel group design. The protocol of the study received approval from the Institutional Review Board of the Sterling Hospital, Ahmedabad. Patients of ACS of either sex who had undergone Percutaneous Transluminal Coronary Angioplasty (PTCA) procedure irrespective of the presence of diabetes mellitus were included in the study after taking their consent. Patients with second or third degree AV (atrioventricular) block, renal or hepatic failure, recent cerebrovascular events, valve replacement surgery, or Balloon Mitral Valvuloplasty (BMV) and patients on lipid lowering therapy other than statins were excluded from the study. 102 consecutive patients meeting the eligibility criteria were randomized into four treatment groups, consisting of patients given atorvastatin, simvastatin, atorvastatin-fenofibrate combination and simvastatin-fenofibrate combination. Treatment was started after the PTCA and was continued for 3 months. Atorvastatin was given in a single dose of 20mg per day (if alone) and 10mg per day (in combination with fenofibrate), simvastatin was given in a single dose of 40mg per day (if alone) and 20mg per day (in combination with fenofibrate) and fenofibrate was given in a single dose of 200mg per day. The first blood samples were collected before beginning of the treatment and were analyzed for total cholesterol, triglycerides(TG), LDL, HDL, VLDL and plasma fibrinogen. The second blood samples were collected after 3 months of treatment for total cholesterol, triglycerides, HDL,

LDL, VLDL, plasma fibrinogen and liver function tests that included SGPT, SGOT, total bilirubin and alkaline phosphatase. Total cholesterol, TG, HDL-C, SGPT, SGOT, total bilirubin and alkaline phosphatase were analyzed on automated VITROS 250 analyzer using enzymatic assay methods. LDL-C was analyzed by enzymatic method. VLDL-C was calculated using Friedewald's formula.

Plasma fibrinogen was assayed using Immunoprecipitation method using the RANDOX® kits. Many studies measure fibrinogen by the Clauss Method. However, data from the Framingham Heart Study [19] suggest that levels determined by the immunoprecipitation test have a stronger association with cardiovascular disease than those obtained by Clauss method. The results were analyzed by applying Student's t test, ANOVA and linear regression to find out degree of correlation between parameters. The probability value of less than 5% ($p<0.05$) was considered to be statistically significant.

Table 1. Baseline Demographic and Haemodynamic characteristics of the Patients

Characteristics	Atorvastatin (n=25)	Atorvastatin + Fenofibrate (n=25)	Simvastatin (n=27)	Simvastatin + Fenofibrate (n=25)
Age(years)	56.76 ± 9.4	56.44 ± 9.95	58.44 ± 11.4	58.35 ± 11.35
Females (%)	4%	12 %	7.4%	20%
Males (%)	96%	88%	92.6 %	80%
Body Mass Index(BMI)	25.79 ± 3.5	23.62 ± 1.95	25.3 ± 3.90	25.44 ± 4.25
Smokers	12%	20%	14.81 %	8%
Tobacco Chewers	12%	12%	7.4 %	4%
Diabetics	24%	36%	29.62%	48%
Hypertensives	28%	28%	51.85 %	40%
Haemoglobin (gm%)	12.61± 1.6	12.17 ± 1.75	13.16 ± 1.76	13.07 ± 1.65
Urea (mg/dl)	30.7± 18.9	26.81 ± 11.95	26.93 ± 15.46	26.91 ± 13.2
Creatinine (mg/dl)	1.23± 0.5	1.28 ± 1.1	1.10 ± 0.31	1.10 ± 0.3
Random bl. Sugar (mg/dl)	130.33± 39.25	146.73± 65.45	137.8 ± 57.45	137.48 ± 56.6
Systolic BP (mm Hg)	126.54 ± 16.05	129.52 ± 24.25	128.22 ± 22.47	128.84 ± 22.8
Diastolic BP (mm Hg)	81.29 ± 8.75	78.35 ± 9.00	80.55 ± 11.31	80.96 ± 11.55
Pulse	77.09 ± 15.4	80.55 ± 10.9	75.02 ± 15.46	75.29 ± 15.65
LVEF (%)	46.8 ± 12.35	50.94 ± 16.25	52.44 ± 12.79	52.51 ± 5.76
Single Vessel disease	68%	56%	55.55%	44%
Double vessel disease	28%	28%	29.62 %	36%
Triple vessel disease	4%	16%	14.81%	20%

Data expressed as mean ± SD; % of patients.

RESULTS

We included in our study, 102 patients from the cardiology unit of Sterling hospital. All of them had angiographically documented CAD and had undergone PTCA. 25 patients each were enrolled in atorvastatin, atorvastatin-fenofibrate combination and simvastatin-fenofibrate combination treatment groups and 27 patients were enrolled in simvastatin treatment group. We had 4 dropouts in atorvastatin-fenofibrate group as well as in simvastatin monotherapy group and 3 dropouts in simvastatin-fenofibrate group, all due to non-technical reasons. So, the analysis is based on a total of 91 patients who completed the follow-up successfully. The baseline, demographic characteristics like age, sex, BMI, smoking/tobacco/ alcohol habits, past history of diabetes, hypertension, family history etc. and other hemodynamic parameters like hemoglobin, urea, creatinine, etc were recorded (Table 1). These parameters were found to be identical in all the four treatment groups indicating a symmetric study design and population.

Table 2. Effect of treatments of serum lipoprotein levels

Parameters	Time interval	Atorvastatin[1] (n=25)	Atorvastatin + Fenofibrate [2] (n=21)	Simvastatin[3] (n=23)	Simvastatin + fenofibrate [4] (n=22)
		Mean ± SD	Mean ± SD	Mean ± SD	Mean ± SD
Total cholesterol mg/dl	baseline	156.6 ± 34.15	150.9± 46.39	150.44 ± 31.1	148.9 ± 31.98
	3 months	149.68 ± 42.8	155.76± 25.42	146.42±53.65	139.74 ± 30.67
TG mg/dl	baseline	144.85±76.65	155.1±112.12	138.5 ± 78.55	136.43 ± 78.97
	3 months	133.86±61.85	108.5±47.63*	123.5±100.7	110.12±47.5
HDL-C mg/dl	baseline	37.69 ±9.1	35.34 ± 9.25	34.91 ± 8.57	34.88 ± 8.67
	3 months	41.55 ±9.2*	46.61±8.38*	43.74 ± 12.9*	43.74 ± 13.27*
LDL-C mg/dl	baseline	101.44 ± 33.5	91.11 ± 33.06	93.39± 23.33	92.10 ± 24.10
	3 months	82.49 ±34.5*	86.74 ± 18.64	76.48 ± 26.3*	76.48 ± 26.35*
VLDL-C mg/dl	baseline	28.96 ±15.2	31.0 ± 22.39	27.68 ± 15.71	27.78 ± 15.85
	3 months	26.87 ± 12.35	21.71±9.66*	24.27 ± 19.97	24.17 ± 20.02
LDL- C/HDL-C	baseline	2.87 ± 1.35	2.66 ± 0.82	2.89 ± 1.05	3.13 ± 1.92
	3 months	2.06 ±0.95*	1.91 ± 0.50*	1.92 ± 0.81*	1.92 ± 0.79*
total cholesterol/ HDL-C	baseline	4.33 ± 1.25	4.41 ± 1.42	4.42 ± 1.29	4.37 ± 1.31
	3 months	3.71 ± 1.2 *	3.42 ± 0.69	3.66 ± 1.77*	3.66 ± 1.78*

Baseline: before starting of treatment;
3 months: after 3 months of treatment;
* Significant change from baseline values; student's t test, $p<0.05$;
$df^1=24$, $df^2=20$, $df^3=22$, $df^4=21$ (degree of freedom).

The effects of all the treatments on lipid parameters are given in Table 2. Atorvastatin and simvastatin monotherapy reduced serum LDL-C and interestingly increased serum HDL-C significantly ($p<0.05$). Atorvastatin-fenofibrate combination produced a significant decrease in TG, VLDL ($p<0.05$) and also increased HDL-C significantly as compared to atorvastatin alone ($p<0.05$). Similar to simvastatin monotherapy, simvastatin-fenofibrate combination produced a significant increase in HDL and decrease in LDL levels ($p<0.05$). It, however, did not reduce TG significantly. All the four treatments significantly reduced total cholesterol/HDL ratio and LDL/HDL ratio ($p<0.05$). ANOVA test was applied to find if there was any difference in the effects between the four groups. The results however, indicated that the percentage change in lipoprotein levels obtained were not significantly different from each other between the four treatment groups (Figure 1).

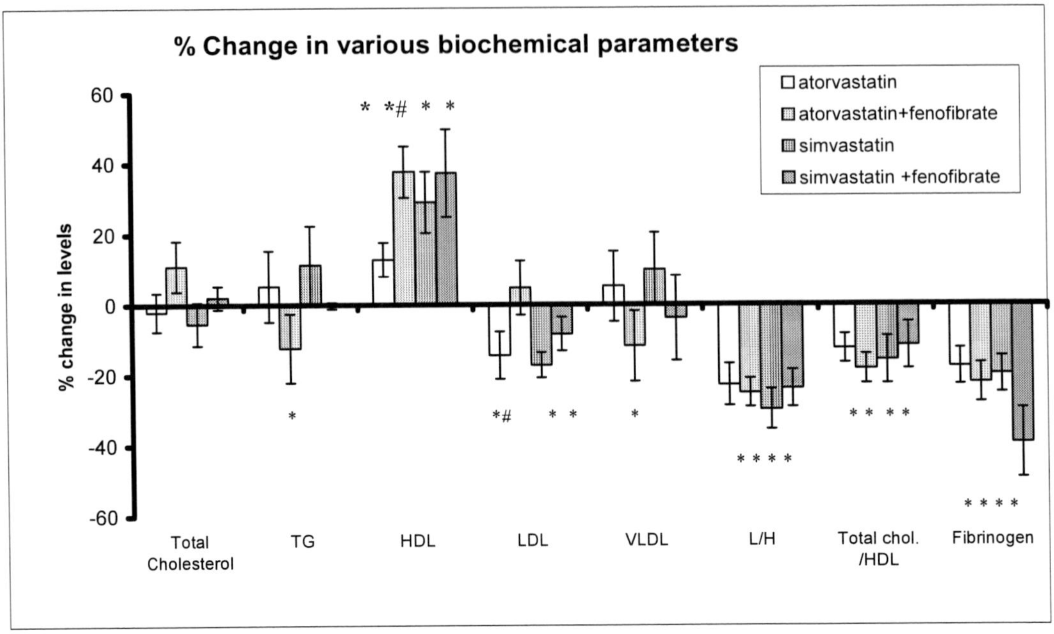

Figure 1. % change in lipid levels between both the treatment group. *Significant change from baseline values, Student's t test, p<0.05. # Significant change as compared to the respective statin monotherapy or combination, Unpaired Student's t test, p<0.05. ANOVA test applied on % change in levels, P>0.05, Changes are not significantly different from each other between the groups.

The effect of the treatments on plasma fibrinogen are given in Table 3. Atorvastatin as well as simvastatin was found to decrease plasma fibrinogen significantly (p<0.01). Combination of statins with fenofibrate also showed a significant decrease in plasma fibrinogen (p<0.01). However, the treatment groups were not significantly different from each other when analyzed by ANOVA test. Thus, contrary to reports, statins in our study decreased plasma fibrinogen significantly while addition of fenofibrate enhanced this effect, however significantly.

Table 3. Effect of treatments on plasma fibrinogen levels

Plasma Fibrinogen Gm/Lt	Atorvastatin [1] (n=25)	Atorvastatin + Fenofibrate [2] (n=21)	Simvastatin[3] (n=23)	Simvastatin + fenofibrate [4] (n=22)
Initial Levels	4.44 ± 0.26	4.29 ± 0.4	4.25 ± 0.25	4.21 ± 0.27
After 3 months	3.47 ±0.18 *	3.10 ± 0.28*	3.12 ± 0.14*	3.12 ± 0.15 *

*significant change from baseline values; Paired Student's t test, p<0.05
$df^1 = 24$, $df^2 = 20$, $df^3 = 22$, $df^4 = 21$.

No correlation was found between fibrinogen levels and various demographical, haemodynamic and biochemical parameters (Table 4). Thus, plasma fibrinogen was found to be an independent risk factor for CAD and was lowered significantly by atorvastatin and simvastatin alone and in combination with fenofibrate.

No significant elevations in the serum bilirubin concentration, transaminases (SGPT, SGOT) and alkaline phosphatase levels were observed in any treatment group (Table 5). It thus appeared that all the treatments were safely tolerated in the given doses without any adverse hepatotoxic effects.

Table 4. Correlation coefficient between fibrinogen and other study parameters

Parameters	Correlation coefficient (r)
Body Mass Index (BMI)	-0.02
Haemoglobin	-0.21
Urea	0.18
Creatinine	-0.21
Blood Glucose	-0.14
Cholesterol	0.05
Triglycerides	0.01
HDL-C	-0.10
LDL-C	0.07
VLDL-C	-0.02
LDL-C/HDL-C	0.10
CHO/HDL-C	0.10

Table 5. Effect of treatments of Liver Function Parameters at the end of three months

Parameters	Atorvastatin[1] (n=25)	Atorvastatin + Fenofibrate [2] (n=21)	Simvastatin[3] (n=23)	Simvastatin + fenofibrate [4] (n=22)
		Mean ± SD	Mean ± SD	Mean ± SD
Total bilirubin mg/dl	0.7 ± 0.25	0.67 ± 0.18	0.71 ± 0.28	0.71 ± 0.28
SGPT U/Lt	28.16 ± 10.9	29.62 ± 7.19	31.67 ± 7.99	31.48 ± 1.72
SGOT U/Lt	24.46 ±7.1	29.14 ± 6.59	27.62 ± 7.28	27.55 ± 1.56
Alk.Phosph-atase U/Lt	94.04 ±23.05	69.00 ± 16.48	74.51 ± 26.25	73.80 ± 5.42

Data expressed as Mean ± SD;
*Significant change from baseline values; student's t test, $p<0.05$;
$df^1=24$, $df^2=20$, $df^3=22$, $df^4=21$ (degree of freedom).

DISCUSSION

Statins alone as well as in combination with fenofibrate were found to produce beneficial effects in CAD patients after PTCA. Statins are potent inhibitors of HMG Co-A reductase, which decreases LDL-C by upregulating LDL receptor activity in the liver and reducing the secretion of apolipoprotein B containing lipoprotein. The latter is believed to be responsible for the TG lowering effect of atorvastatin and is profound at higher doses.[20] Various reports have indicated that atorvastatin lowers LDL-C and increases HDL.[21,22,23] The effects of atorvastatin in our study comply with such reports. On the other hand, simvastatin is reported to produce a decrease in LDL-C from 26% to 50% in various doses.[24, 25] Also, reports from Hunninghake et al [26] showed that simvastatin consistently produces a larger increase in HDL-C as compared to atorvastatin. Results from our study comply with the above findings as simvastatin in our study too produced a significant decrease in LDL-C and increase in HDL-C as compared to atorvastatin.

Fibrates activate PPAR-α which ultimately leads to their HDL raising and TG lowering effects.[27,28] Certain previous studies have suggested increased benefits of a combination of fenofibrate with statins. [29,30] Reports are available that show benefits of combining fenofibrate and simvastatin. [30, 31] A study of the literature also shows that a greater change in HDL and TG levels is obtained with fenofibrate - atorvastatin combination as compared to monotherapy [32] with either drug. Our results comply with reported findings. We observed that combining fenofibrate with statins offered a greater benefit in reducing TG and VLDL levels as well as increasing HDL levels. Thus, the combination therapy proved both safe and beneficial in our patients with ACS.

In prospective studies, plasma fibrinogen has been found to be an independent predictor of myocardial infarction in both genders. It provides information on CAD risk over and beyond that supplied by established risk factors. [33] The Framingham study has shown that for both sexes, the risk of cardiovascular diseases was correlated positively to antecedent fibrinogen values higher than 1.3 to 7.0 gm/L.[34] We found that fibrinogen in our study did not correlate with any of our study parameters including demographic characters and lipid levels. Thus, fibrinogen does appear to be an independent risk factor for CAD.

Reports from large-scale trials consistently show statins to have a neutral effect on fibrinogen. [35,36] Various studies on plasma fibrinogen indicate that fenofibrate lowers fibrinogen levels but the effects of atorvastatin and simvastatin have been variable and controversial, particularly that of atorvastatin. Wierzbicki et al [37] showed that atorvastatin increases plasma fibrinogen by 22%. Song et al [38] too found a significant increase in fibrinogen with atorvastatin treatment where as simvastatin was reported to have a neutral effect. Various reports indicated that fenofibrate decreased plasma fibrinogen significantly whereas atorvastatin produced an increase in fibrinogen levels.[39, 40,41] Otto et al [42] showed that plasma fibrinogen and other hemorheologic parameters were unchanged during atorvastatin treatment in comparison to simvastatin treatment. Athyros et al [43] reported that plasma fibrinogen was unaffected by atorvastatin and was significantly reduced by fenofibrate and combination of both. Ceska et al [44] showed that fenofibrate is a potent hypolipidemic drug with only rare side effects and reduces fibrinogen significantly. In our study, however, contrary to all above reports for statins, we observed that atorvastatin as well as simvastatin produced a significant decrease in plasma fibrinogen levels. However, our results comply with some recent findings by Kadikoylu and co- workers that show a decrease in plasma fibrinogen by atorvastatin and simvastatin. [45] Another study by Leibovitz et al [47] proved that atorvastatin reduces fibrinogen levels in patients with severe hypercholesterolemia. fibrinogen levels dropped by almost 18% in this study. Tekin et al [48] has also shown a reduction in plasma fibrinogen levels by atorvastatin in hyperlipidemic patients with angiographically proved CAD. These findings with atorvastatin are consistent with such recent data. The combination of statins with fenofibrate also produced a further decrease in fibrinogen, though not significant. These results are in agreement with the reported claims of fenofibrate to decrease plasma fibrinogen.

Previous studies have indicated an increased risk of myopathy with statin-fibrate combination. [29,46] However, no cessation of therapy was required in any patient due to such complication in our study. Liver function tests of all patients were also normal at the end of 3 months of treatments indicating no adverse hepatotoxic effects. These findings suggest that it may be safe to use fenofibrate and statins in combination contrary to reported contra-indications for the combined use of these two classes of drugs. However, we have not looked into the long-term toxicity associated with the use of these drugs in combination. Moreover, the strategy appears to be distinctly beneficial in lowering the risk factor plasma fibrinogen. Thus, the therapy offers a new method to treat patients with acute coronary syndromes.

Conclusions

Addition of fenofibrate to statins seems to be beneficial in patients with ACS. Statins, contrary to various reports, were found to decrease plasma fibrinogen significantly. Also, in combination with fenofibrate there was enhanced reduction of the novel risk factor, fibrinogen.

References

[1] Murray JL, Lopez AL. Alternative projections of mortality and disability by cause 1990-2020. Global Burden of Disease Study. *Lancet* 1997; 349:1498-1504.

[2] Tsimihodimos V, Kostoula A, Kakafika A, Bairaktari E, Tselepis AD, Mikhailidis DP, Elisaf M. Effect of fenofibrate on serum inflammatory markers in patients with high triglyceride values. *J Cardiovasc Pharmacol Ther* 2004; 9:27-33.

[3] Kockx M, Gervois PP, Poulain P, Derudas B, Peters JM, Gonzalez FJ, Princen HM, Kooistra T, Staels B. Fibrates suppress fibrinogen gene expression in rodents via activation of the peroxisome proliferators-activated receptor-alpha. *Blood* 1999; 93: 2991-8.

[4] Green F, Humphries S. Control of plasma fibrinogen levels. *Baillieres Clin Haematol* 1989;2: 945-59.

[5] Woods A, Brull DJ, Humphries SE, Montgomery HE. Genetics of inflammation and risk of coronary artery disease: the central role of interleukin-6. *Eur Heart J* 2002; 21: 1574-83.

[6] Shepherd, J., Cobbe, S. M., Ford, I. Prevention of coronary heart disease with pravastatin in men with hypercholesterolemia. *N Engl J Med* 1995; 333: 1301-7.

[7] Sacks, F. M., Pfeffer, M. A., Moye, L. A. The effect of pravastatin on coronary events after myocardial infarction in patients with average cholesterol levels. *N Engl J Med* 1996 335:1001-9

[8] The Scandinavian Simvastatin Survival Group. Randomized trial of cholesterol lowering in 4444 patients with coronary heart disease: The Scandinavian Simvastatin Survival Study (4S). *Lancet* 1994; 344: 1383-9.

[9] Athyros, V. G., Papageorgiou, A. A., Hatzikonstandinou, H. A., Athyrou, V. V., Kontopoulos, A. G.Effect of atorvastatin versus simvastatin on lipid profile and plasma fibrinogen in patients with hypercholesterolemia. *Clin Drug Inves* 1998; 16: 219-27.

[10] Rosenson RS, Tangney CC, Casey LC. Inhibition of proinflammatory cytokine production by pravastatin. *Lancet* 1999; 353: 983-4.

[11] Rosenson RS, Tangney CC. Anti-atherothrombotic properties of statins: implications for cardiovascular event reduction. *JAMA* 1998; 279:1643-50.

[12] Handley DA, Hughes TE. Pharmacological approaches and strategies for therapeutic modulation of fibrinogen. *Thromb Res* 1997: 87:1-5.

[13] Ernst E, Resch KL. Therapeutic interventions to lower plasma fibrinogen concentration. *Eur Heart J* 1995; 16:47-51.

[14] Fuller GM. Fibrinogen: a multifunctional acute phase protein, in Mackiewicz A, Kushner I, Baumann H (eds): Acute phase proteins: Molecular Biology, biochemistry and clinical applications, New York, NY, Doubleday, 1993,p 169.

[15] Kant JA, Fornace AJ, Saxe D, Simon MI, McBride OW, Crabtree GR. Evolution and organization of the fibrinogen locus on chromosome 4: Gene duplication accompanied by transposition and inversion. *Proc Natl Acad Sci* USA 1985: 82: 2344-50

[16] Binsack R, Stegmeier K, Dorge L, Volkl A. Bezafibrate down-regulates fibrinogen biosynthesis in human hepatoma HepG2 cells. *Eur J Clin Invest* 1998; 28:151-5.

[17] Roy SN, Mukhopadhyay G, Redman CM. Regulation of fibrinogen assembly. Transfection of Hep G2 cells with B beta cDNA specifically enhances synthesis of the 3 component chains of fibrinogen. *J Biol Chem* 1990; 265:6389-93.

[18] Maison, P., Mennon, L., Sapinho, D., Balkau, B., Sigalas, J., Chesnier, M. C., Eschwege, E. D.E.S.I.R. Study group. A pharmacoepidemiological assessment of the effect of statins and fibrates on fibrinogen concentration. *Atheroscler* 2002; 160:155-60

[19] Stec JJ, Silbershatz H, Tofler GH, Matheney TH, Sutherland P,Lipinska I, Massaro JM, Wilson PF, Muller JE, D'Agostino RB Sr. Association of fibrinogen with cardiovascular risk factors and cardiovascular disease in the Framingham Offspring Population. *Circ* 2000; 102: 1634-8.

[20] Burnett JR, Wilcox LT, Telford DE, Kleinstiver SJ, Barrett PHR, Newton RS, Huff MW. Inhibition of HMG Co A reductase by atorvastatin decreases both VLDL and LDL, apoB productionin miniature pigs. *Arterioscler Thromb Vasc Biol* 1997;17:2489-2500.

[21] Haw N, Fowler G, Patel H, Eminton Z, Maton S. An assessment of the efficacy of atorvastatin in achieving LDL cholesterol target levels in patients with coronary heart disease: a general practice study. *Int J Clin Prac* 1999;53:422-6.

[22] Marais AD, Firth JC, Bateman M, Jones J, Mountney J. Atorvastatin is a powerful and safe agent for lowering plasma cholesterol concentration in heterozygous familial hypercholesterolemia. *Atheroscler* 1994;109:316-20.

[23] Frost R, Otto C, Geiss C, Schwandt P, Parhofer KG. Effects of atorvastatin versus fenofibrate on lipoprotein profiles, low density lipoprotein subfraction distribution, and

hemorheologic parameters in type 2 diabetes mellitus with mixed hyperlipoproteinemia. *Am J Cardiol* 2001;87:44-8.

[24] Jones, P., Kafonek, S., Laurova, I. Comparative dose efficacy study of atorvastatin versus simvastatin, pravastatin, lovastatin and fluvastatin in patients with hypercholesterolemia. (The CURVES Study). *Am J Cardiol* 1998; 81: 582-7.

[25] Crouse JR III, Frolich J, Ose L, Mercuri M, Tobert JA. Effects of high doses of simvastatin and atorvastatin on high-density lipoprotein cholesterol and apolipoprotein A-1. *Am J Cardiol* 1999;83:1476-7.

[26] Hunninghake DB, Ballantyne CM, Maccubbin DL, Shah AK, Gumbiner B, Mitchel YB. Comparative effects of simvastatin and atorvastatin in hypercholesterolemic patients with characteristics of metabolic syndrome. *Clin Ther* 2003;25:1670-86.

[27] Staels B, Dallongvile J, Auwerx J, Schoonjans K, Leitersdorf E, Fruchart JC. Mechanism of action of fibrates on lipid and lipoprotein metabolism. *Circ* 1998;98:2088-93.

[28] Watts GF, Dimmit SB. Fibrates, dyslipoproteinemia and cardiovascular disease. *Curr Opin Lipid* 1999;10:561-74.

[29] Athyros VG, Papageorgiou AA, Hatziknonstandinou HA, Didangelos TP, Carina MV, Kranitsas DF, Kontupoulos AG. Safety and efficacy of long term statin-fibrate combination in patients with refractory familial combined hyperlipidemia. *Amer J Cardiol* 1997; 80: 608-13.

[30] Ruth LBE and Ruth McPherson. Long term efficacy and safety of fenofibrate and statin in treatment of combined hyperlipidemia. *Amer J Cardiol* 1998; 81: 60B-65B.

[31] Vega GL, Ma PT, Cater NB, Fillipchuk N, Megura S, Garcia AB, Grundy SM. Effects of adding fenofibrate (200mg/day) in patients with combined hyperlipidemia and metabolic syndrome. *Amer J Cardiol* 2003; 91: 956-60.

[32] Keating GM, Ormod D. Micronised fenofibrate: an updated review of its clinical efficacy in the management of dyslipidaemia. *Drugs* 2002; 62:1909-44.

[33] Heinrich, J., Assmann, G. Fibrinogen and cardiovascular risk. *J Cardiovasc Risk* 1995;2: 197-205.

[34] Kannel, W. B., Neaton, J. D., Wentworth, D., Thomas, H. E., Stamler, J., Hulley, S. B. For the MRFIT Research group. Overall and coronary heart disease mortality rates in relation to major risk factors in 325,348 men screened for the MRFIT trial. Multiple Risk Factor Intervention Trial. *Am Heart J* 1987; 112:825-36.

[35] Rosenson RS, Koenig W. Utility of inflammatory markers in the management of Coronary artery disease. *Amer J Cardiol* 2003; 92:10i-18i.

[36] Lowe G, Rumley A, Norrie J, Ford I, Shepherd J, Cobbe S, Macfarlane P, Packard C, on behalf of the West of Scotland Coronary Prevention Study Group. Blood Rheology, cardiovascular risk factors and cardiovascular disease: the west of Scotland Coronary Prevention Study. *Thromb Haemost* 2000; 84: 553-8.

[37] Wierzbicki, A. S., Lumb, P. J., Semra, Y. K. Effect of atorvastatin on plasma fibrinogen. *Lancet* 1998; 351:569-70.

[38] Song, J. C., White, C. M. Do HMG CoA reductase inhibitors affect fibrinogen? *Ann Pharmacother* 2001; 35:236-41.

[39] Bairaktari, E. T., Tzallas, C. S., Tsimihodimos, V. K., Liberopoulos, E. N., Miltiadous, G. A., Elisaf, M. S. Comparison of the efficacy of atorvastatin and micronized fenofibrate in the treatment of mixed hyperlipidemia. *J Cardiovasc Risk* 1999; 6:113-6.

[40] Frost, R., Otto, C., Geiss, C., Schwandt, P., Parhofer, K. G. Effects of atorvastatin versus fenofibrate on lipoprotein profiles, low density lipoprotein subfraction distribution, and hemorheologic parameters in type 2 diabetes mellitus with mixed hyperlipoproteinemia. *Am J Cardiol* 2001; 87:44-8.

[41] Serna, G., Cadarso, C. Fenofibrate decreases plasma fibrinogen, improves lipid profile, and reduces uricemia. *Clin Pharmacol Ther* 1999; 66:166-72.

[42] Otto, C., Geiss, H. C., Donner, M. G., Parhofer, K. G., Schwandt, P. Influence of atorvastatin versus simvastatin on fibrinogen and other hemorheological parameters in patients with severe hypercholesterolemia treated with regular low density lipoprotein immunoadsorption apheresis. *Ther Apher* 2000; 4:244-8.

[43] Athyros, V. G., Papageorgiou, A. A., Athyrou, V. V., Demitriadis, D. S., Kontopoulos, A. G. Atorvastatin and micronized fenofibrate alone and in combination in type 2 diabetes with combined hyperlipidemia. *Diab Care* 2002; 25:1198-1202.

[44] Ceska R, Sobra J, Kvasnicka J, Prochazkova R, Kvasilova M, Haas T. The effect of micronized fenofibrate on lipid parameters and fibrinogen in heterozygous familial hypercholesterolemia and familial combined hyperlipidemia. *Cas Lek Cesk* 1996;135: 413-6.

[45] Kadikoylu, G., Yukselen, V., Yavasoglu, I., Bolaman, Z. Hemostatic effects of atorvastatin versus simvastatin. *Ann Pharmacother* 2003; 37:478-84.

[46] Tikkanen MJ. Statins within-group comparisons, statin escape and combination therapy. *Curr Opin Lipidol* 1996; 7: 385-8.

[47] Leibovitz E, Hazanov N, Frieman A, Elly I, Gavish D. Atorvastatin reduces fibrinogen levels in patients with severe hypercholesterolemia: additional evidence to support the anti-inflammatory effects of statins. *Isr Med Assoc J* 2004;6:456-9.

[48] Tekin A, Tekin G, Guzelsoy D, Kaya A, Gurel CV, Yigit Z, Ulutin T. Effects of atorvastatin (10 mg) on hemostatic and inflammatory parameters in hyperlipidemic patients with angiographically proven coronary artery disease. *Am J Cardiol.* 2004;94:206-9.

In: Statins: Indications and Uses, Safety and Modes of Action ISBN 978-1-60692-103-6
Editor: Gunnar N. Holmquist

Chapter 5

NON-LIPID LOWERING EFFECTS OF STATINS

*A. Schmidt**
Institute for Arteriosclerosis Research, Domagkstr. 3, University of Muenster, D-48149 Muenster, Germany

ABSTRACT

The beneficial effect of statins on the reduction of cardiovascular events can be only partly attributed to their cholesterol lowering effect. The antiproliferative, anti-inflammatory, and immunomodulatory properties of statins appear to be largely unrelated to lipid-lowering but may be explained by affecting post-translational modification or isoprenylation essential for membrane localization and biologic activity of several proteins including adapter proteins and enzymes involved in signal transduction pathways. The present data reinforce the hypothesis that statins may represent innovative pharmacological tools not only for the prevention of cardiovascular related disease in normolipidemic patients but also for diseases where a reduced isoprenylation of regulatory proteins reveals benefit effects.

PLEIOTROPIC EFFECTS OF STATINS

HMG-CoA reductase inhibitors (statins) are approved for lowering the blood plasma cholesterol level by inhibiting of 3-hydroxy-3-methylglutaryl-CoA reductase, the rate-limiting enzyme of cholesterol synthesis. Statins have been shown to improve survival of patients with hypercholesterolemia and coronary heart disease, but large clinical trials with statins reveal early benefit effects of cardiac events that are independent of the cholesterol-lowering properties of statins and unrelated to angiographically detected regression. These

* Correspondence concerning this article should be addressed to: PD Dr. rer. nat. Annette Schmidt, Leibniz-Institute for Arteriosclerosis Research Domagkstr. 3 D-48149 Muenster, Germany. Phone: + 49-(0)251-835 8626; Fax: + 49-(0)251-835 8628; E-mail: annschm@uni-muenster.de.

results suggest that statins have atheroprotective effects and further pharmacological properties in addition to reducing the plasma LDL level.

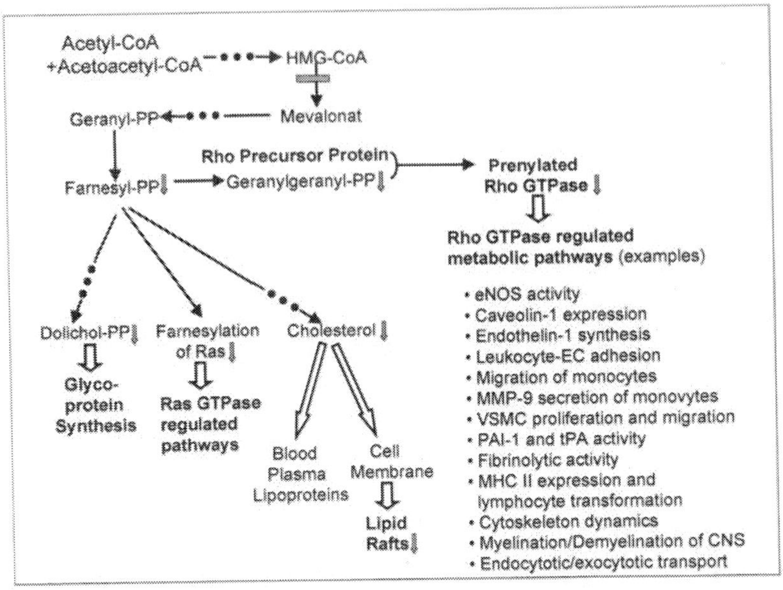

Figure 1. Inhibitory effects of statins on the mevalonate pathway and related sideways. The inhibition of HMG-CoA reductase is limiting not only for the synthesis of cholesterol but also for several intermediates required for posttranslational isoprenylation of proteins and their activation and correct localization in distinct cellular compartments.

The rate limiting step in cholesterol synthesis is the formation of mevalonate from HMG-CoA. The mevalonate pathway then branches out before the synthesis of squalen and cholesterol. These sideways of the mevalonate pathway go out from farnesylpyrophosphate and generate further important polyisoprenoid effectors: dolicholpyrophosphate (involved in glycoprotein synthesis), ubichinon (essential for respiratory chain in mitochondria), farnesylpyrophosphate (required for prenylation of Ras) and geranylgeranylpyrophosphate (required for prenylation of Rho, Rac, Rab, Rap) (Figure 1). Small GTPases of the Rho family play a crucial role in the regulation of cell growth and differentiation, cytoskeleton assembly, cell motility, protein and lipid trafficking, nuclear transport and host defence [1-7]. The posttranslational prenylation of these regulatory proteins creates lipid binding sites that influence the membrane binding and intracellular trafficking and thus the biological activity of these proteins (Figure 2). In fact, small G proteins who's proper membrane localization and function are dependent on isoprenylation are related with the effect of HMG-CoA reductase inhibitors and impaired isoprenoid synthesis. Because protein prenylation depends on a free availability of intermediates of the mevalonate pathway and all isoprenoids belong to the class of lipids the concept "lipid-independent effects of statins" does not appear justified. Moreover, statins do not only lower the blood plasma cholesterol but also the structure and function of cell membranes. For example, the response of lymphocytes to exogenous signals such as antigens is orchestrated by a number of molecules that cluster in cholesterol rich areas of the cell membrane known as lipid rafts. Lipid rafts stabilize the cell membrane and act as platforms bringing together molecules essential for the activation of immune cells but also separating such molecules when the conditions for activation are not appropriate. The statin-mediated depletion of cholesterol synthesis alters stability and fluidity of membranes and could impair the function of lipid rafts. Other pathways affected by statins seem to be the regulation of the activity of the cholesteryl ester transfer protein and their anti-inflammatory effect by binding to the lymphocyte function associated antigen-1 [4].

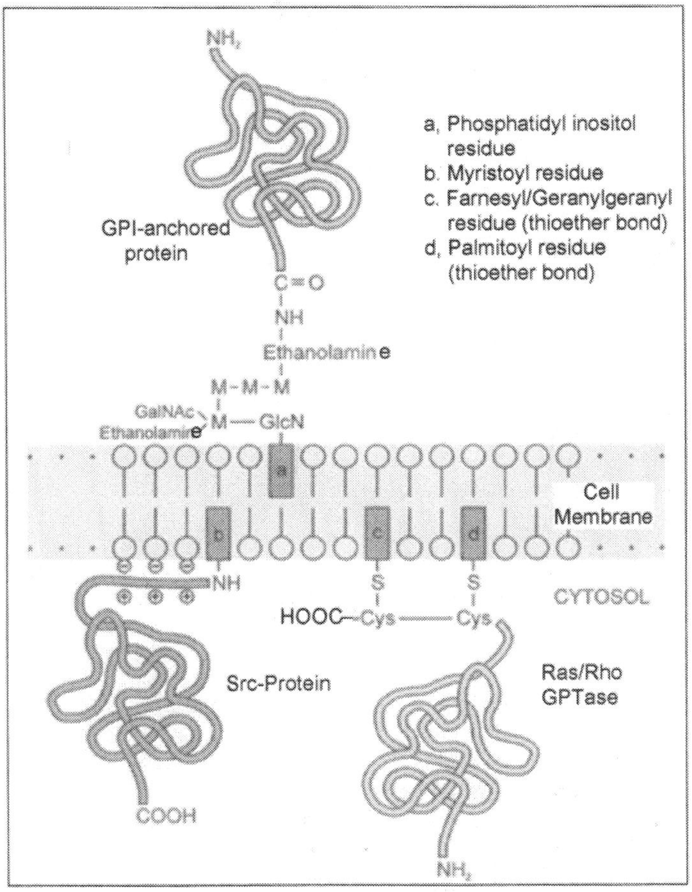

Figure 2. Posttranslational linkage of isoprenoids (and fatty acids) enables proteins to anchor in the cytosolic layer of plasma membranes. The GPI anchor is embedded in the outerpart of the membrane lipid bilayer.

PROLIFERATION, MIGRATION AND APOPTOSIS

As judged from [^{3}H]thymidine incorporation and cell counting, statins caused a reduction of DNA synthesis and proliferation in human vascular smooth muscle cells (vSMC) and human vascular endothelial cells [8,9]. Under *in vitro* conditions statins arrest the cell cycle primarily during the G_1/S-phase [8,10,11]. The proliferation inhibitory effect was associated with a reduced expression of bFGF, a potent mitogen of arterial smooth muscle cells and with a diminished signalling response to VEGF and insulin-like growth factor [12]. *In vitro* studies demonstrated that most statins may decrease vSMC proliferation [13] and migration [14] in a dose-dependent manner. Compared to the significant antiproliferative potency of simvastatin, fluvastatin, lovastatin and atorvastatin a minimal inhibitory effect on vSMC proliferation has been shown for pravastatin *in vitro* [15,16]. These different antiproliferative properties of the statins may be attributed to differences in their lipophilicity and ability to penetrate the cells. The effect of statins on the intracellular signalling showed an inhibition of kDa p44 MAPK and a reduced phosphorylation of p42 MAPK. While the attenuated phosphorylation of p44 MAPK was partly restored by addition of mevalonate, the reduced phosphorylation of p42 MAPK could not be restored by addition of excessive doses of mevalonate or by stimulation of the cells with basic fibroblast growth factor. Concurrently the expression of the GTP-binding Ras protein was significantly elevated at 5 and 20 µmol/L lovastatin, this effect being attenuated by addition of mevalonate to cell cultures [17]. It has been concluded that the modulation of smooth muscle cell metabolism by HMG-CoA reductase inhibitors may be divided into two categories: one well-described type involved in inhibition of cell growth and associated metabolic events [8,17], which can be prevented by supplementation of the cells with mevalonate; the second category is apparently independent of the polyisoprenoid pathway and involves highly specific phosphorylation cascades of the intracellular signalling pathway. Lovastatin could also reduce smooth muscle cell migration. Using modified Boyden chambers it was shown that the invasion of smooth muscle cells towards a chemotactic stimulus (PDGF over 24 hours) was dose-dependently reduced in the presence of 0.5-5 µM simvastatain [18] by reducing isoprenylation of the small GTP binding protein Rho [19]. VEGF is suggested to be involved in atherosclerotic plaque growth by inducing neovascularization. HMG-CoA reductase inhibitors (e.g. atorvastatin) have atheroprotective effects by reduction of VEGF protein that was associated with a significantly attenuated activity of the transcription factors AP-1 and HIF-1 and with a diminished expression of VEGF mRNA [20].

The impact of statins on vSMC apoptosis was shown by the finding that atorvastatin, simvastatin and lovastatin but not pravastatin induced apoptosis in cultured vascular SMC in a dose–dependent manner, an effect that was primarily mediated by the inhibition of protein prenylation [21,22].

Statins Stabilize Atherosclerotic Plaque by Enzyme Inhibition

Plaque resident cells (smooth muscle cells, macrophages) synthesize and secrete matrix metalloproteinases that degrade extracellular matrix components and increase the risk of plaque rupture by lysis of the fibrous cap overlaying the lipid core of the plaque. The matrix metalloproteinases-1,-3 and -9, secreted by rabbit vSMC, were inhibited by lovastatin in a dose-dependent fashion, while the tissue inhibitors of MMPs (TIMP-1 and -2) were not reduced. The inhibition of MMP-9 was due to a failed prenylation of small G-proteins required for MMP-9 gene transcription [23]. Hypercholesterolemia has been demonstrated to be associated with an enhanced cellular production of superoxide anions *in vitro* and *in vivo*.

In patients with polygenic hypercholesterolemia, platelet formation of superoxide anion was significantly higher and significantly related to LDL cholesterol than in sex- and age-matched normal cholesterolemia subjects. An eight-week administration of statins (10 mg atorvastatin/day) significantly reduced LDL cholesterol and platelet production of superoxide anion. This could be explained by the cholesterol lowering effect of the statin only to a certain extent, in that it was observed after 3 days of statin treatment only and was not parallel to cholesterol reduction suggesting that other mechanisms could be responsible for the antioxidant activity of the drug such as inhibition of phospholipase A2 or inhibition of NADH/NADPH oxidase. The activity of both enzymes could be reduced by applying specific inhibitors (AACOCF3, DPI) [24].

Anti-Inflammatory Properties

Inflammation occurs in the vasculature as a response to injury, lipid peroxidation and perhaps infection. Numerous *in vitro* and *in vivo* studies provide evidence for benefit of statins by directly or indirectly modulating the inflammatory component of arteriosclerosis. The interaction between blood leukocytes and the vascular endothelium represents a crucial inflammatory step in the atherogenic process [25]. An expression of cell adhesion molecules (E-selectin, ICAM-1 and VCAM-1) is an essential prerequesit for endothelial transmigration of leukocytes including monocytes as present in hypercholesterolemia. It was shown that the basal release of NO from endothelial cells is associated with an increase in endothelial cell adhesion molecule expression [26-28] and that statins may increase the endothelial nitric oxide synthase activity [29]. In contrast simvastatin was shown to attenuate the cell adhesion molecule expression in apolipoprotein E-deficient mice in a lipid independent manner [30] and to cause a significant reduction of leukocyte endothelial cell interaction in a normocholesterolemic rat. This was at least partly mediated by attenuated upregulation of P-selectin on endothelial cells [31]. The small GTP binding protein Rho is essential for integrin (L-selectin)-depending adhesion of leukocytes to the endothelium. Because geranylgeranylpyrophosphate is required for the activation of Rho it has been concluded that the cell adhesion molecule modulating effect of statins results in part by inhibition of geranylgeranylation of Rho. In the same way fluvastatin was shown to inhibit the interaction between monocytes and human umbilical vein endothelial cells by lowering the expression of ICAM-1 on monocytes [32] and simvastatin was shown to attenuate the CD18 upregulation in polymorph nuclear leukocytes in response to stimulation with leukotrien B4 in normocholesterolemic rat [33]. Cerivastatin was reported to reduce monocyte adhesion to vascular endothelium by downregulation of integrin adhesion molecules (CD11a, CD18) and inhibition of actin polymerization via prevention of Rho translocation to the membrane [34].

Another anti-inflammatory effect of statins is the reduced production of proinflammatory cytokines. Fluvastatin and pravastatin were found to inhibit the secretion of interleukin-6 in response to angiotensin II in cultured human vSMC and fluvastatin and simvastatin but not pravastatin reduced the production of interleukin-6 and interleukin-1b in human umbilical vein endothelial cells [35]. Pravastatin and cerivastatin have been shown to inhibit the expression of MCP-1 and transforming growth factor $ß_1$ associated with inhibition of eNOS (endothelial nitric oxide synthase) in rat hearts [36]. Atorvastatin and pravastatin produced a similar reduction of MCP-1 expression in different arterial SMC [37].

Simvastatin inhibits the inflammatory properties of *staphylococcus aureus* alpha-toxin. This has been shown in a study where after pretreatment with simvastatin (50-100 µg/kg) the mesenteric microcirculation was induced by a bolus administration of 40 µg/kg *staphylococcus aureus* alpha-toxin resulting in a significant and time-dependent increase in leukocyte rolling, adherence and transmigration of leukocytes as well as P-selectin expression in the intestinal vascular endothelium. Pretreatment with simvastatin significantly inhibited exotoxin-induced alterations and was associated with an enhanced expression of endothelial cell NO synthase-3 [38].

Increased levels of C-reactive protein (CRP), the classic acute phase reactant and a sensitive marker for inflammation, could be used as an effective marker for cardiovascular risk. A reduced rate of progression of arteriosclerosis associated with intensive statin treatment was significantly related to a greater reduction of the levels of CRP regardless of the resultant level of LDL-cholesterol [39,40] indicating that the anti-inflammatory potential of statins is mediated to a considerable part independent of cholesterol lowering. Although the magnitude of risk reduction associated with statin use appears to be largest for patients with the highest serum levels of CRP whether CRP reduction per se lowers cardiovascular risk is unknown [41].

Immunomodulating Properties

The clinical relevant immune effects of statins concern:

(a) inhibition of the expression of class II major histocompatibility antigens (MHC-II), and
(b) blocking of the ß-2 integrin leukocyte function.

(a) Statins inhibit the expression of MHC-II on human macrophages, endothelial cells and smooth muscle cells stimulated by interferon gamma [2,42]. This effect was exerted by both lipophilic and hydrophilic statins at nM to µM concentrations, but constitutively MHC-II expressing cells (B-cells, dendritic cells) were not affected. The inhibition of the expression of MHC-II was due to an inhibition of the promoter of the MHC-II transactivator (CIITA) which regulates transcription and synthesis of MHC-II. Then MHC-II inhibition leads to a reduction of C-cell proliferation and differentiation while Th1 cells secrete cytokines that promote inflammation such as IFN-gamma (interferon-gamma) [43], Th2 cells inhibit Th1 cell proliferation by a negative feedback mechanism. Probably the overall effect of reduced T-cell activation and proliferation would be beneficial.

(b) Statins are selectively blocking the beta-2 integrin leukocyte function antigen-1 (LFA-1) [44]. LFA-1 (CD11a/CD18) is expressed on the surface of leukocytes and binds in its activated form to ICAM-1. In addition it is a costimulator of T-cells. On analysis of healthy volunteers atorvastatin (20 mg) lead to a significant downregulation of HLA-DR and

CD38 activation marker on peripheral T-cells, whereas simvastatin upregulated both of these molecules. In contrast T-cell activation was inhibited by simvastatin and enhanced by atorvastatin. These immunomodulatory effects of statins on human T-cells demonstrated *in vivo* a different effect of two statins. Atorvastatin lead to a major histocompatibility class II antigen downregulation and might therefore be investigated for treatment of chronic transplant rejection, simvastatin inhibited superantigen mediated T-cell activation which might explain reduced mortality of simvastatin treated patients with staphylococcal bacteriemia [45]. A potential molecular target of statin is the promoter 4 for the transactivator CIITA that is specific for inducible MHC-II expression and does not concern constitutive expression of CIITA and MHC-II [46].

Statins have also immunomodulating effects involving a reduction in hemodynamically significant rejection episodes of cardiac transplantation recipients (pravastatin) [47], an inhibition of TNF-α and inducible iNOS expression by microglia and astrocytes (lovastatin) [48]. In the central nervous system atorvastatin promoted differentiation of Th0 cells into Th2 cells by inducing STAT6 phosphorylation and secretion of Th2 cytokines (IL-4, IL-5, IL-10) and TGF-ß. Conversely, STAT4 phosphorylation was inhibited and secretion of Th1 cytokines (IL-2, IL-12, interferon-gamma, and TNF-α) was suppressed. These effects were associated with a prevention or reversion of chronic and relapsing experimental encephalomyelitis, a demyelinating disease model of multiple sclerosis [49]. Using a mouse air-pouch model of local inflammation in peripheral blood mononuclear cells exposed to lipopolysaccharides and in cultured human endothelial cells exposed to IL-1ß, lovastatin and pravastatin reduced leukocyte recruitment and monocyte chemotactic protein-1 (MCP-1) production [50].

The beneficial effect of statins (pravastatin, simvastatin) in plaque inflammation and stability can be shown in cynomolgus monkey under an atherogenic diet for 12 months. On statin treatment, the macrophage content of plaques, the expression of VCAM-1, IL-1ß, and tissue factor were about 2-fold lowered, while the collagen content was 1.5 - 2.1-fold greater versus control animals [51].

Statin-Mediated Promotion of Metabolic Pathways

Inhibition of HMG-CoA reductase results in a reduced intracellular production of mevalonate not only as an essential precursor of cholesterol, but also of several isoprenoids required for translocation of prenylated proteins to membranes for proper activity. However, lacking prenylation of proteins could also affect inhibitors of metabolic pathways. This explains the observation that statins (lovastatin, simvastatin) cause a superinduction and overexpression of endothelial E-selectin, ICAM-1 and VCAM-1 [52,53] possibly by involvement of a prenylated inhibitory G protein. Statins also upregulate the expression of nitric oxide synthesis [54]. Furthermore, in endothelial cells and endothelial precursor cells statins (0.5 μM simvastatin) rapidly activate the protein kinase Akt/PKB that promotes endothelial survival, nitric oxide production, and differentiation *in vitro*. The resulting cellular responses contribute to new blood vessel growth and stabilization of the vascular network. The promotion of the translocation of Akt to discrete sites in endothelial cell plasma membrane was shown to be a PI3-kinase-dependent process in migratory endothelial cells [55].

Interleukin-8 (IL-8) is characterized as pro-inflammatory cytokine that plays an important role as a mitogen and chemoattractant for vSMC [56]. Vascular smooth muscle cell migration can be stimulated by IL-8 to values 20-fold over those of controls, a crucial step in the development of vessel wall thickening during atherosclerosis. In human coronary smooth muscle cells statins (lovastatin or simvastatin) markedly increased the intracellular IL-8 levels up to 23-fold [57] in contrast to a decreased production of IL-8 in human THP-1 monocytes under lovastatin treatment [58]. If comparable results can be found under *in vivo* conditions in the vessel wall the known clinical benefits of statins could not be explained by primarily lowering the pro-inflammatory cytokine IL-8 in the arterial wall.

Statins Prevent LDL-Proteoglycan Interaction

Statin-exposed vSMC secrete proteoglycans with decreased binding affinity for LDL. Monkey aortic smooth muscle cells grown in culture were exposed to simvastatin (10-100 μM) and cerivastatin (0.1 and 1 μM) and newly secreted proteoglycans were quantified and characterized [9]. Both, simvastatin and cerivastatin caused a concentration-dependent reduction in cell growth and reduced $^{35}SO_4$ incorporation into secreted proteoglycans on both an absolute and per cell basis. The decrease in $^{35}SO_4$ incorporation is in part explained by decreased cellular proliferation but $^{35}SO_4$ incorporation is also reduced on a per cell basis. The finding, that statins do not decrease the relative ^{35}S-methionine incorporation into proteoglycan core proteins suggest that there is no change in the molar concentration of proteoglycans per cell. Therefore statin exposure appears to decrease the degree of sulfation of glycosaminoglycan side chains. In addition to being undersulfated proteoglycans secreted by statin-exposed cells have a larger apparent molecular weight and hydrodynamic size mediated by an elongation of glycosaminoglycan side chains. The undersulfated less negatively charged proteoglycans from statin-exposed cells bind less LDL than do proteoglycans secreted from control cells. Decreasing the binding interaction of LDL and the subsequent reduced retention of LDL in the arterial intima may benefit arteriosclerosis in a manner unrelated to serum cholesterol lowering [59]. Furthermore, cell membrane-associated heparan sulfate proteoglycan has been reported to bind and internalize aggregated LDL in a receptor-independent manner [60]. Proteoheparan sulfate adsorbed to a methylated (hydrophobic) silica surface in a monolayer has a high binding affinity to LDL at normal blood Ca^{2+} concentration. This ternary aggregation of heparan sulfate, LDL, and Ca^{2+} results in the formation of surface-bound “nanoplaques” that *in vivo* could build up a nucleus for increasing plaque size. Fluvastatin, whether applied to the patient (one single dose of 80 mg) or in *in vitro* experiments (2.2 μmol/L), markedly slowed down the process of plaque formation. This immediate effect of fluvastatin occurred without any change in lipid concentration of the patient [61,62].

STATINS AND CANCER

Statins have been shown to inhibit proliferation and to induce apoptosis in a variety of tumor cells [63]. They have also been found to display antitumor effects against melanoma, mammary carcinoma, pancreatic adenocarcinoma, fibrosarcoma, glioma, neuroblastoma and lymphoma in animal tumor models resulting in retardation of tumor growth and/or inhibition of the metastatic process [63-65]. The molecular mechanisms underlaying antitumor activity of statins have not been fully elucidated but interference with the function of Ras and Rho-family GTPases, inhibition of the activity of cyclin-dependent kinases and activation of cyclin-dependent kinase inhibitors could contribute to this activity [63].

In a study on human breast cancer cells [66] it was demonstrated that cerivastatin reduces the proliferation and invasion of aggressive breast cancer cells. The molecular mechanism could be explained by the cerivastatin effect on gene expression (microarray) and signal transducing pathways. The expression of 13 genes was modified by cerivastatin and confirmed at protein level. The downregulation of cyclin D1, PCNA, c-myc and upregulation of p21 (Waf-1), p19 (INK-4d), integrin beta-8 could explain the inhibition of cell proliferation and cell invasion either directly (decrease in uPA, MMP-9, uPAR, PAI-1 and increase in anti-oncogenes Wnt-5a and H-cadherin) or indirectly by stimulating an antiangiogenic gene (thrombospondin-2). The antiangiogenic activity was confirmed by *in vivo* experiments. Furthermore it was demonstrated that the biochemical mechanisms of its anticancer action could mainly be explained by the inhibition of RhoA-dependent cell signalling. This hypothesis was supported by the fact that a RhoA inhibitor (C3 exoenzyme) or a dominant negative mutant RhoA (N19-RhoA) induced similar effects to those of cerivastatin. It was concluded that cerivastatin by preventing RhoA prenylation inhibits the RhoA kinase pathway leading to defective actin stress fibre formation responsible for the loss of traction forces required for cell mobility and the RhoA/Fak/Akt signalling pathway that could explain the majority of cancer related gene modifications described above. Thus the inhibition of RhoA cell signalling could be a good strategy in therapy for aggressive forms of breast cancer [66].

HEMOSTASIS AND THROMBOGENICITY

Antithrombotic activity of the statins (simvastatin, fluvastatin and cerivastatin) was mediated by reducing the expression of tissue factor (TF) by cultured human monocytes/macrophages in a dose-dependent manner, an effect that was reversed by coincubation with mevalonate or geranylgeranylpyrophosphate. Moreover, fluvastatin-induced inhibition of bacterial lipopolysaccharide-stimulated tissue factor gene expression in human monocytes/macrophages was associated with an inhibition of the activation of NF-kappaB [67,68]. Simvastatin was shown to inhibit the expression of plasminogen activator inhibitor-1 (PAI-1) from human vSMC and endothelial cells while increasing the expression of tissue type plasminogen activator from endothelial cells [69]. This effect was due to the inhibition of Rho geranylgeranylation and disruption of cellular actin filaments [70]. Pravastatin has been found to prolong the clotting time *in vitro*. In combination with low molecular weight heparin pravastatin resulted in a significantly prolonged clotting time compared with pravastatin and low molecular weight heparin given alone. The synergistic effect of pravastatin and low molecular weight heparin on prolongation of the clotting time suggests that pravastatin exerts its effect by inhibition of the coagulation cascade and fibrin formation [71].

Many studies have suggested variable mechanisms by which statins may exert an antithrombotic effect that include a significant downregulation of tissue factor expression followed by a reduced thrombin generation and corresponding impairment of several thrombin-dependent coagulant reactions (fibrinogen/fibrin, factor V/Va, factor XIII/XIIIa). Moreover, a statin-triggered increase of thrombomodulin expression on endothelial cells may enhance the activity of protein C anticoagulant pathway. Most of the antithrombotic effects of statins (atorvastatin, simvastatin, fluvastatin, lovastatin) are attributed to the inhibition of isoprenylation of signalling molecules [6]. Pravastatin has been observed to have a different effect on fibrinogen levels in several studies and has been demonstrated to have inhibitory effects on platelet activation [72-76].

STATINS AND RHEUMATOID ARTHRITIS

Patients with rheumatoid arthritis (RA) and systemic Lupus erythematodes (SLE) have a significant increased risk of cardiovascular disease and therefore might benefit from statin therapy although it can not be excluded that statins also improve the autoimmune aspect. The list of disorders for which statins may prove beneficial is growing and now extends from multiple sclerosis and neurodegenerative disorders to RA and SLE. Among the first reports of the immunological effects of statins was the finding that this class of drugs inhibits the increase in cell surface protein of major histocompatibility complex class II induced by interferon gamma. The MHC II class are central in presenting antigen and activating T-cells and their expression is often increased in inflammation. Increased production of interferon gamma by activated T-cell is characteristic of a number of autoimmune disease in human including collagen-induced arthritis and experimental autoimmune encephalomyelitis, a murine model of multiple sclerosis [77]. Another aspect of the statin-mediated reducing the inflammatory component of cardiovascular disease is the association between cholesterol levels and immunological regulation. Cholesterol is the key component of the structure and raft forming function of cell membranes. Thus, the T-cell receptor (TCR) and costimulatory molecules including lymphocyte function associated antigen–I (LFA-I), CD28, CD4 and CD40 ligand (CD40 L) are recruited to lipid rafts after activation (Figure 3). Statins interfere with the activation of T-cells by depleting membrane cholesterol and disruption the integrity of lipid rafts. Statin treatment causes the exclusion from lipid microdomain of raft-associated molecules such as the LCK protein tyrosine kinase, the inhibition of actin polymerisation and the formation of a stable immunological synapse and therefore disruption of T-cell activation [77]. RA synovitis contains a predominant Th1 response, widespread macrophage, fibroblast, mast cell and B-cell activation that in turn generate high autoantibody production and a cytokine release (TNF-alpha, Interleukin-1ß, -6, -15 and –18). Endothelial cell activation, upregulation of adhesion molecule expression and angiogenesis are increasingly recognized.

Simvastatin has been shown *in vitro* to suppress synovial T-cell and fibroblast-like synoviocytes cytokines release [78]. Extented studies show that simvastatin effectively reduce the severity of rhodent collagen-induced arthritis *in vivo* [79]. In a randomised, double blind, placebo controlled trial, it was found that atorvastatin (40 mg/day) suppressed acute phase parameters (CRP 50%) and significantly reduced the swollen joint in patients with RA presenting with active disease despite existing disease modifying antirheumatic drug treatment [80].

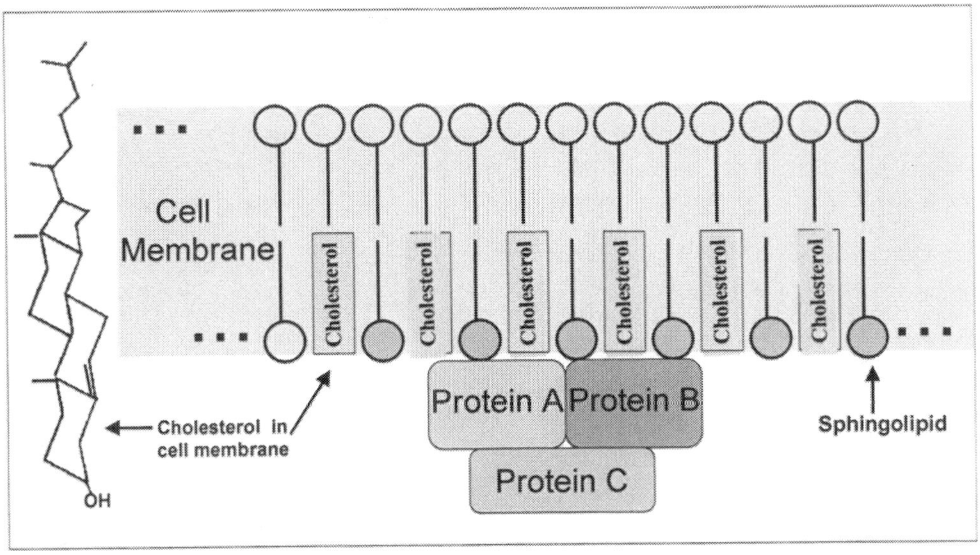

Figure 3. Statins cause alterations in the expression and distribution of lipid raft domains in cell membranes. Lipid rafts act as platform bringing together molecules essential for activation of signalling pathways. In T-cells receptors and costimulatory molecules are recruited to lipid rafts [77].

STATINS AS ANTI-ARRHYTHMIC THERAPY

It has been proposed that statins reduce the incidence of arrhythmias in patients with atherosclerotic heart disease [81-83]. Thus the anti-inflammatory effect of statins can reduce the recurrence of artrial fibrillation (AF) in patients who undergo successful cardioconversion [82]. This effect is in accordance with the hypothesis that inflammation, evidenced by high levels of C-reactive protein can induce artrial fibrillation and promote its persistence. It was found that after two years of follow up patients treated with statin therapy for high cholesterol had a significantly lower artrial fibrillation recurrence rate than the patients non on lipid lowering therapy (40% vs 84%, p=0.007). Beside this indirect antiarrhythmic effect statins could excert direct antiarrhythmic effects by modulating the lipid composition and physicochemical properties of cell membranes with a resultant alteration in transmembrane ion channel properties [84,85].

Multiple Sclerosis

Statins decreased the migration of leukocytes into the central nervous system, inhibited MHC class II and costimulatory signals required for activation of proinflammatory T-cells induced a Th2 phenotype in T-cells and decreased the expression of inflammatory mediators in the central nervous system including nitric oxide and tumor necrosis factor alpha [86].

On the basis of the known immunomodulatory effects of statins a multicentre, open label, single arm study to gather information on use of oral simvastatin (80 mg) was designed. In 30 individuals with relapsing remitting multiple sclerosis the mean number of gadolinium enhancing lesions at month 4, 5 and 6 of treatment was compared with the mean number of lesions noted on pretreatment brain MRI scans. Number and volume of gadolinium enhanced lesions declined by 44%, ($p<0.0001$) and 41% ($p=0.0018$) respectively. Treatment was well tolerated. These findings suggest that an 80 mg daily dose of oral simvastatin over a 6 month period could inhibit the inflammatory components of multiple sclerosis that lead to neurological disability. Since individuals were enrolled on the basis of their disease activity on baseline MRI scans noted reductions in a baseline versus treatment trial design could represent regression to the mean [86]. An open label clinical trial of simvastatin for multiple sclerosis confirmed the result of these studies revealing a significant decrease in the number and volume of new MRI lesions and a favourable safety profile [41].

Conclusions

The beneficial effect of statins on the reduction of cardiovascular events can be only partly attributed to their cholesterol lowering effect. The antiproliferative, anti-inflammatory, and immunomodulatory properties of statins appear to be largely unrelated to lipid-lowering but may be explained by affecting post-translational modification or isoprenylation essential for membrane localization and biologic activity of several proteins including adapter proteins and enzymes involved in signal transduction pathways. The present data reinforce the hypothesis that statins may represent innovative pharmacological tools not only for the prevention of cardiovascular related disease in normolipidemic patients but also for diseases where a reduced isoprenylation of regulatory proteins reveals benefit effects.

References

[1] Sherer, Y; Shoenfeld, Y. Immunomodulation for treatment and prevention of atherosclerosis. *Autoimmunity Rev,* 2002;1:21-7.

[2] Palinski, W; Tsimikas, S. Immunomodulatory Effects of Statins: Mechanisms and Potential Impact on Arteriosclerosis. *J Am Soc Nephrol*, 2002;13:1673-81.

[3] Palinski, W. New Evidence for Beneficial Effects of Statins Unrelated to Lipid Lowering. *Arterioscler Thromb Vasc Biol*, 2001;21:3-5.

[4] Bonetti, PO; Lerman, LO; Napoli, C; Lerman A. Statin effects beyond lipid lowering – are they clinically relevant? *Eur Heart J*, 2003;24:225-48.

[5] Mach, F. Statins as Immunomodulatory Agents. *Circulation*, 2004;108 (Suppl II):II-15-II-17.

[6] Undas, A ; Brummel-Ziedins, KE ; Mann, KG. Statins and Blood Coagulation. *Arterioscler Thromb Vasc Biol*, 2005;25:287-94.

[7] Raja, SG; Dreyfus, GD. Statins: Much More than Just A Lipid-Lowering Therapy. *Indian Heart J*, 2004;56:204-9.

[8] Sindermann, JR; Fan, L; Weigel, KA; Troyer, D; Muller, JG; Schmidt, A; March, KL; Breithardt, G. Differences in the effects of HMG-CoA reductase inhibitors on proliferation and viability of smooth muscle cells in culture. *Atherosclerosis*, 2000;150:331-41.

[9] Meyers, CD; Tannock, LR; Wight, TN; Chait, A. Statin-exposed vascular smooth muscle cells secrete proteoglycans with decreased binding affinity for LDL. *J Lipid Res*, 2003;44:2152-60.

[10] Jakobisiak, M; Bruno, S; Skierski, JS; Darzynkiewicz, Z. Cell cycle-specific effects of lovastatin. *Proc Natl Acad Sci USA*, 1991;88:3628-32.

[11] Skaletz-Rorowski, A; Eschert, H; Pawlus, E; Breithardt G. Molecular effects of HMG-CoA reductase inhibitors on smooth muscle cell proliferation. *J Am Coll Card*, 2001;337-38 (Letter).

[12] Vincent, TS; Wulfert, E; Merler, E. Inhibition of growth factor signalling pathways by lovastatin. *Biochem Biophys Res Commun*, 1991;180:1284-89.

[13] Axel, DI; Rissen, R; Runge, H; Viebahn, R; Karsch, KR. Effects of cerivastatin on human arterial smooth muscle cell proliferation and migration in transfilter cocultures. *J Cardiovasc Pharmacol*, 2000;35:619-29.

[14] Corsini, A; Arnaboldi, L; Raiteri, M; Quarato, P; Faggiotto, A; Paoletti, R; Fumagalli, R. Effect of the new HMG-CoA reductase inhibitor cerivastatin (BAY W 6228) on migration, proliferation and cholesterol synthesis in arterial myocytes. *Pharmacol Res*, 1996;33:55-61.

[15] Nègre-Aminou, P; van Vliet, AK; van Erck, M; van Thiel, GC; van Leeuwen, RE; Cohen, LH. Inhibition of proliferation of human smooth muscle cells by various HMG-CoA reductase inhibitors; comparison with other human cell types. *Biochim Biophys Acta*, 1997;1345:259-68.

[16] Corsini, A; Mazzotti, M; Raiteri, M; Soma, MR; Gabbiani, G; Fumagalli, R; Paoletti, R. Relationship between mevalonate pathway and arterial myocyte proliferation: in vitro studies with inhibitors of HMG-CoA reductase. *Atherosclerosis*, 1993;101:117-25.

[17] Sindermann, JR; Schmidt, A; Breithardt, G; Buddecke E. Lovastatin controls signal transduction in vSMC by modulating phosphorylation levels of mevalonate-independent pathways. *Basic Res Cardiol*, 2001;96:283-89.

[18] Porter, KE; Naik, J; Turner, NA; Dickinson, T; Thompson, MM; London, NJ. Simvastatin inhibits human saphenous vein neointima formation via inhibition of smooth muscle cell proliferation and migration. *J Vasc Surg*, 2002;36:150-57.

[19] Laufs, U; Marra, D; Node,K; Liao, JK. 3-hydroxy-3-methylglutaryl-CoA reductase inhibitors attenuate vascular smooth muscle cell proliferation by preventing Rho GTP ase-induced down-regulation of p27^{Kip1}. *J Biol Chem*, 1999;274:21926-31.

[20] Alber, HF; Dulak, J; Frick, M; Dichtl, W; Schwarzacher, SP; Pachinger, O; Weidinger, F. Atorvastatin decreases vascular endothelial growth factor in patients with coronary artery disease. *J Am Coll Cardiol*, 2002;39(12):1951-5.

[21] Stark, WW Jr; Blaskovich, MA; Johnson, BA; Qian, Y; Vasudevan, A; Pitt, B; Hamilton, AD; Sebti, SM; Davies, P. Inhibiting geranylgeranylation blocks growth and promotes apoptosis in pulmonary vascular smooth muscle cells. *Am J Physiol*, 1998;275:L55-63.

[22] Guijarro, C; Blanco-Colio, LM; Ortego, M; Alonso, C; Ortiz, A; Plaza, JJ; Diaz, C; Hernandez, G; Egido, J. 3-hydroxy-3-methylglutaryl coenzyme A reductase and isoprenylation inhibitors induce apoptosis of vascular smooth muscle cells in culture. *Circ Res*, 1998;83:490-500.

[23] Luan, Z; Chase, AJ; Newby, AC. Statins inhibit secretion of metalloproteinases-1, -2, -3, and -9 from vascular smooth muscle cells and macrophages. *Arterioscler Thromb Vasc Biol*, 2003;23(5):769-75.

[24] Sanguigni, V; Pignatelli, P; Caccese, D; Pulcinelli, FM; Lenti, L; Magnaterra, R; Martini, F; Lauro, R; Violi, F. Increased superoxide anion production by platelets in hypercholesterolemic patients. *Thromb Haemost*, 2002;87:796-801.

[25] Ross, R. The pathogenesis of atherosclerosis: an update. *N Engl J Med*, 1986;314:488-500.

[26] Gauthier, TW; Scalia, R; Murohara, T; Guo, JP; Lefer, AM. Nitric oxide protectcs against leukocyte-endothelium interactions in the early stages of hypercholesterolemia. *Arterioscler Thromb Vasc Biol*, 1995;15:1652-9.

[27] Davenpeck, KL; Gauthier, TW; Lefer, AM. Inhibition of endothelial-derived nitric oxide promotes P-selectin expression and actions in the rat microcirculation. *Gastroenterology*, 1994;107:1050-8.

[28] Lefer, AM; Ma, XL. Decreased basal nitric oxide release in hypercholesterolemia increases neutrophil adherence to rabbit coronary artery endothelium. *Arterioscler Thromb*, 1993;13:771-6.

[29] Laufs, U; La Fata, V; Plutzky, J; Liao, JK. Upregulation of endothelial nitric oxide synthase by HMG-CoA reductase inhibitors. *Circulation*, 1998;97:1129-35.

[30] Scalia, R; Gooszen, ME; Jones, SP; Hoffmeyer, M; Rimmer, DM 3rd; Trocha, SD; Huang, PL; Smith, MB; Lefer, AM; Lefer, DJ. Simvastatin exerts both anti-inflammatory and cardioprotective effects in apolipoprotein E-deficient mice. *Circulation*, 2001;103:2598-603.

[31] Preufer, D; Scalia, R; Lefer, AM. Simvastatin inhibits leukocyte-endothelial interactions and protects against inflammatory processes in normocholesterolemic rats. *Arterioscler Thromb Vasc Biol,* 1999;19:2894-900.

[32] Niwa, S; Totsuka, T; Hayashi, S. Inhibitory effect of fluvastatin, an HMG-CoA reductase inhibitor, on the expression of adhesion molecules on human monocyte cell line. *Int J Immunopharmacol*, 1996 ;18:669-75.

[33] Lefer, AM ; Campbell, B ; Shin, YK ; Scalia, R; Hayward, R; Lefer, DJ. Simvastatin preserves the ischemic-reperfused myocardium in normocholesterolemic rat hearts. *Circulation*, 1999;100:178-84.

[34] Yoshida, M; Sawada, T; Ishii, H. HMG-CoA reductase inhibitor modulates monocyte-endothelial cell interaction under physiological flow conditions in vitro: involvement of Rho GTPase-dependent mechanism. *Arterioscler Thromb Vasc Biol,* 2001;21:1165-71.

[35] Inoue, I; Goto, S; Mizotani, K; Awata, T; Mastunaga, T; Kawai, S; Nakajima, T; Hokari, S; Komoda, T; Katayama, S. Lipophilic HMG-CoA reductase inhibitor has an anti-inflammatory effect: Reduction of mRNA levels for interleukin-1ß, interleukin-6, cyclooxygenase-2, and p22phox by regulation of peroxisome proliferator-activated receptor alpha (PPARalpha) in primary endothelial cells. *Life Sci*, 2000;67:863-76.

[36] Ni, W; Egashira, K; Kataoka, C; Kitamoto, S; Koyanagi, M; Inoue, S; Takeshita, A. Antiinflammatory and antiarteriosclerotic actions of HMG-CoA reductase inhibitors in a rat model of chronic inhibition of nitric oxide synthesis. *Circ Res*, 2001;89:415-21.

[37] Martinez-González, J; Alfón, J ; Berrozpe, M ; Badimon, L. HMG-CoA reductase inhibitors reduce vascular monocyte chemotactic protein-1 expression in early lesions from hypercholesterolemic swine independently of their effect on plasma cholesterol levels. *Atherosclerosis*, 2001;159:27-33.

[38] Pruefer, D; Makowski, J; Schnell, M; Buerke, U; Dahm, M; Oelert, H; Sibelius, U; Grandel, U; Grimminger, F; Seeger, W; Meyer, J; Darius, H; Buerke, M. Simvastatin inhibits inflammatory porperties of Staphylococcus aureus alpha-toxin. *Circulation*, 2002;106(16):2104-10.

[39] Ridker, PM; Cannon, CP; Morrow, D; Rifai, N; Rose, LM; McCabe, CH; Pfeffer, MA; Braunwald, E. Pravastatin or Atorvastatin Evaluation and Infection Therapy-Thrombolysis in Myocardial Infarction 22 (PROVE IT-TIMII 22) Investigators. C-reactive protein levels and outcomes after statin therapy. *N Eng J Med*, 2005;352(1):20-8.

[40] Nissen, SE; Tuzcu, EM; Schoenhagen, P; Crowe, T; Sasiela, WJ; Tsai, J; Orazem, J; Magorien, RD; O'Shaughnessy, C; Ganz, P. Reversal of Atherosclerosis with Aggressive Lipid Lowering (REVERSAL) Investigators. Statin therapy, LDL cholesterol, C-reactive protein, and coronary artery disease. *N Engl J Med*, 2005;352(1):29-38.

[41] Neuhaus, O; Stuve, O; Zamvil, SS; Hartung, HP. Are statins a treatment option for multiple sclerosis? *Lancet Neurol*, 2004;3(6):369-71.

[42] Kwak, B; Mulhaupt, F; Myit, S; Mach, F. Statins as a newly recognized type of immunomodulator. *Nat Med*, 2000;6:1399-1402.

[43] Palinski, W. Immunomodulation: A new role for statins ? *Nat Med*, 2000;6:1311-2.

[44] Weitz-Schmidt, G; Welzenbach, K; Brinkmann, V; Kamata, T; Kallen, J; Bruns, C; Cottens, S; Takada, Y; Hommel, U. Statins selectively inhibit leukocyte function antigen-1 by binding to a novel regulatory integrin site. *Nat Med*, 2001;7:687-92.

[45] Fehr, T; Kahlert, C; Fierz, W; Joller-Jemelka, HI; Riesen, WF; Rickli, H; Wuthrich, RP; Ammann, P. Statin-induced immunomodulatory effects on human T cells in vivo. *Atherosclerosis*, 2004;175(1):83-90.

[46] Mach, F. Immunosuppressive effects of statins. *Atheroscler Suppl*, 2002;3(1):17-20.

[47] Kobashigawa, JA; Katznelson, S; Laks, H; Johnson, JA; Yeatman, L; Wang, XM; Chia, D; Terasaki, PI; Sabad, A; Cogert, GA; et al. Effect of pravastatin on outcomes after cardiac transplantation. *N Engl J Med*, 1995;333:621-27.

[48] Wong, B; Lumma, WC; Smith, AM; Sisko, JT; Wright, SD; Cai, TQ. Statins suppress THP-1 cell migration and secretion of matrix metalloproteinase-9 by inhibiting geranylgeranylation. *J Leukoc Biol*, 2001;69:959-62.

[49] Youssef, S; Stuve, O; Patarroyo, JC; Ruiz, PJ; Radosevich, JL; Hur, EM; Bravo, M; Mitchell, DJ; Sobel, RA; Steinman, L; Zamvil, SS. The HMG-CoA reductase inhibitor, atorvastatin, promotes a Th2 bias and reverses paralysis in central nervous system autoimmune disease. *Nature*, 2002;420:78-84.

[50] Romano, M; Diomede, L; Sironi, M; Massimiliano, L; Sottocorno, M; Polentarutti, N; Guglielmotti, A; Albani, D; Bruno, A; Fruscella, P; Salmona, M; Vecchi, A; Pinza, M; Mantovani, A. Inhibition of monocyte chemotactic protein-1 synthesis by statins. *Lab Invest*, 2000;80:1095-100.

[51] Sukhova, GK; Williams, JK; Libby P. Statins reduce inflammation in atheroma of nonhuman primates independent of effects on serum cholesterol. *Arterioscler Thromb Vasc Biol*, 2002;22:1452-58.

[52] Schmidt, A; Goepfert, Ch; Feitsma, K; Buddecke E. Lovastatin-stimulated superinduction of E-selectin, ICAM-1 and VCAM-1 in TNF-alphą-activated human vascular endothelial cells. *Atherosclerosis*, 2002;164:57-64.

[53] Sadeghi, MM; Collinge, M; Pardi, R; Bender, JR. Simvastatin modulates cytokine-mediated endothelial cell adhesion molecule induction: Involvement of an inhibitory G protein. *J Immunol*, 2000;165:2712-18.

[54] Laufs, U; La Fata, V; Plutzky, J; Liao, JK. Upregulation of endothelial nitric oxide synthase by HMG-CoA reductase inhibitors. *Circulation*, 1998;97:1129-35.

[55] Skaletz-Rorowski, A; Lutchman, M; Kureishi, Y; Lefer, DJ; Faust, JR; Walsh, K. HMG-CoA reductase inhibitors promote cholesterol-dependent Akt/PKB translocation to membrane domains in endothelial cells. *Cardiovasc Res*, 2003;57(1):253-64.

[56] Yue, TL; Wang, X; Sung, CP; Olson, B; McKenna, PJ; Gu, JL; Feuerstein, GZ. Interleukin-8. A mitogen and chemoattractant for vascular smooth muscle cells. *Circ Res*, 1994;75:1-7.

[57] Sindermann, JR; Schmidt, A; Breithardt, G. Upregulation of the interleukin-8 sys tem in hypercholesterolemic patients. Does inhibition of the mevalonate pathway lower interleukin-8 levels in the vessel wall? *Atherosclerosis*, 2000;150:443-44.

[58] Terkeltaub, R; Solan, J; Barry, M Jr; Santoro, D; Bokoch, GM. Role of the mevalonate pathway of isoprenoid synthesis in IL-8 generation by activated amoncytic cells. *J Leukocyte Biol*, 1994;55:749-55.

[59] Davignon, J; Laaksonen, R. Low-density lipoprotein-independent effects of statins. *Curr Opin Lipidol*, 1999;10:543-59.

[60] Llorente-Cortes, V; Otero-Vinas, M; Badimon, L. Differential role of heparan sulfate proteoglycans on aggregated LDL uptake in human vascular smooth muscle cells and mouse embryonic fibroblasts. *Arterioscler Thromb Vasc Biol*, 2002;22:1905-11.

[61] Abletshauser, C; Klussendorf, D; Schmidt, A; Winkler, K; Marz, W; Buddecke, E; Malmsten, M; Siegel, G. Biosensing of arteriosclerotic nanoplaque formation and

interaction with an HMG-CoA reductase inhibitor. *Acta Physiol Scand*, 2002;176:131-45.

[62] Siegel, G; Abletshauser, C; Malmsten, M; Schmidt, A; Winkler K. Reduction of arteriosclerotic nanoplaque formation and size by fluvastatin in a receptor-based biosensor model. *Cardiovasc Res*, 2003;58:696-705.

[63] Jakobisiak, M; Golab, J. Potential antitumor effects of statins. *Int J Oncol*, 2003;23:1055-69.

[64] Cauley, JA; Zmuda, JM; Lui, LY; Hillier, TA; Ness, RB; Stone, KL; Cummings, SR; Bauer, DC. Lipid-lowering drug use and brast cancer in older women: a prospective study. *J Womens Health (Larchmt)*, 2003;12:749-56.

[65] Collisson, EA; Kleer, C; Wu, M; De, A; Gambhir, SS; Merajver, SD; Kolodney, MS. Atorvastatin prevents RhoC isoprenylation, invasion, and metastasis in human melanoma cells. *Mol Cancer Ther*, 2003;2:941-8.

[66] Denoyelle, C; Albanese, P; Uzan, G; Hong, L; Vannier, JP; Soria, J; Soria, C. Molecular mechanism of the anti-cancer activity of cerivastatin, an inhibitor of HMG-CoA reductase, on aggressive human breast cancer cells. *Cell Signal*, 2003;15(3):327-38

[67] Colli, S; Eligini, S; Lalli, M; Camera, M; Paoletti, R; Tremoli, E. Vastatins inhibit tissue factor in cultured human macrophages: a novel mechanism of protection against atherothrombosis. *Arterioscler Thromb Vasc Biol,* 1997;17:265-72.

[68] De Martin, R; Hoeth, M; Hofer-Warbinek, R; Schmidt, JA. The transcription factor NF-kappaB and the regulation of vascular cell function. *Arterioscler Thromb Vasc Biol,* 2000;20:E83-8.

[69] Bourcier, T; Libby, P. HMG-CoA reductase inhibitors reduce plasminogen activator inhibitor-1 expression by human vascular smooth muscle and endothelial cells. *Arterioscler Thromb Vasc Biol,* 2000;20(2):556-62.

[70] Essig, M; Nguyen, G; Prié, D; Escoubet, B; Sraer, JD; Friedlander, G. 3-hydroxy-3-methylglutaryl coenzyme A reductase inhibitors increase fibrinolytic activity in rat aortic endothelial cells. Role of geranylgeranylation and Rho proteins. *Circ Res*, 1998;83:683-90.

[71] Zimmer, JE; Spillert, CR; Puppala, S; Zamecki, K; Bhatt, BA; Arora, RR. Pravastatin potentiates the anticoagulant effects of low molecular weight heparin. *Thromb Res*, 2004;113(6):407-10.

[72] Zimmer, JE ; Spillert, CR ; Puppala, S ; Zamecki, K ; Bhatt, BA ; Arora, RR. Pravastatin potentiates the anticoagulant effects of low molecular weight heparin. *Thromb Res*, 2004;113:407-10.

[73] Dangas, G; Badimon, JJ; Smith, DA; Unger, AH; Levine, D; Shao, JH; Meraj, P; Fier, C; Fallon, JT; Ambrose, JA. Pravastatin therapy in hyperlipidemia: effects on thrombus formation and the systemic hemostatic profile. *J Am Coll Cardiol*, 1999;33:1294-304.

[74] Dangas, G; Smith, DA; Unger, AH; Shao, JH; Meraj, P; Fier, C; Cohen, AM; Fallon, JT; Badimon, JJ; Ambrose, JA. Pravastatin: an antithrombotic effect independent of the cholesterol-lowering effect. *Thromb Haemost*, 2000;83:688-92.

[75] Tsuda, Y; Satoh, K; Kitadai, M; Takahashi, T; Izumi, Y; Hosomi, N. Effects of pravastatin sodium and simvastatin on plasma fibrinogen level and blood rheology in type II hyperlipoproteinemia. *Atherosclerosis*, 1996;122:225-33.

[76] Branchi, A; Rovellini, A; Sommariva, D; Gugliandolo, AG; Fasoli, A. Effect of three fibrate derivatives and of two HMG-CoA reductase inhibitors on plasma fibrinogen level in patients with primary hypercholesterolemia. *Thromb Haemost*, 1993;70:241-3.

[77] Ehrenstein, MR; Jury, EC; Mauri, C. Statins for Atherosclerosis- As Good as It Gets ? *N Eng J Med,* 2005;352(1):73-5.

[78] McInnes, IB; McCarey, DW; Sattar, N. Do statins offer therapeutic potential in inflammatory arthritis? *Ann Rheum Dis*, 2004;63:1535-7.

[79] Leung, BP; Sattar, N; Crilly, A; Prach, M; McCarey, DW; Payne, H; Madhok, R; Campbell, C; Gracie, JA; Liew, FY; McInnes, IB. A novel anti-inflammatory role for simvastatin in inflammatory arthritis. *J Immunol*, 2003;170:1524-30.

[80] McCarey, DW; McInnes, IB; Madhok, R; Hampson, R; Scherbakov, O; Ford, I; Capell, HA; Sattar, N. Trial of Atorvastatin in Rheumatoid Arthritis (TARA): a double-blind randomised placebo-controlled trial. *Lancet*, 2004;363:2015-21.

[81] Mitchell, LB; Powell, JL; Gillis, AM; Kehl, V; Hallstrom, AP; AVID Investigators. Are lipid-lowering drugs also antiarrhythmic drugs? An analysis of the Antiarrhythmics versus Implantable Defibrillators (AVID) trial. *J Am Coll Cardiol*, 2003;42:81-7.

[82] Siu, CW; Lau, CP; Tse, HF. Prevention of atrial fibrillation recurrence by statin therapy in patients with lone atrial fibrillation after successful cardioversion. *Am J Cardiol*, 2003;92:1343-5.

[83] Young-Xu, Y; Jabbour, S; Goldberg, R; Blatt, CM; Grayboys, T; Bilchik, B; Ravid, S. Usefulness of statin drugs in protecting against atrial fibrillation in patients with coronary artery disease. *Am J Cardiol*, 2003;92:1379-83.

[84] Lamers, JM; Hartog, JM; Verdouw, PD; Hulsmann, WC. Dietary fatty acids and myocardial function. *Basic Res Cardiol*, 1987;82(Suppl I):209-221.

[85] Pound, EM; Kang, JX, Leaf, A. Partitioning of polyunsaturated fatty acids, which prevent cardiac arrhythmias, into phospholipid cell membranes. *J Lipid Res*, 2001;42:346-51.

[86] Vollmer, T; Key, L; Durkalski, V; Tyor, W; Corboy, J; Markovic-Plese, S; Preiningerova, J; Rizzo, M; Singh, I. Oral simvastatin treatment in relapsing-remitting multiple sclerosis. *Lancet*, 2004;363:1607-08.

In: Statins: Indications and Uses, Safety and Modes of Action ISBN 978-1-60692-103-6
Editor: Gunnar N. Holmquist

Chapter 6

STATINS IN THE THERAPY OF ACUTE CORONARY SYNDROME

Petr Ostadal[1,*] and Jan Vojacek[2]

[1]Department of Cardiology, Cardiocenter, Third Faculty of Medicine, Charles University in Prague and University Hospital Kralovske Vinohrady, Prague, Czech Republic

[2]Department of Cardiac Surgery, Second Faculty of Medicine, Charles University in Prague and University Hospital Motol, Prague, Czech Republic

ABSTRACT

It has been repeatedly shown that statins (3-hydroxy-3-methylglutaryl coenzyme A reductase inhibitors) are very effective in primary and secondary prevention of ischemic heart disease. In the settings of acute coronary syndrome (ACS), different pathological pathways are triggered that are known to be inhibited by statins, including endothelial dysfunction and activation of inflammation and coagulation; the idea to use these drugs also under conditions of ACS seems to be, therefore, fully justified. Recently, several prospective controlled clinical trials have been presented, showing safety and in some points also efficacy of statins, when administered early after ACS. An increasing number of publications demonstrates, however, that statins may express a positive effect not only in the early secondary prevention but also directly in the therapy of ACS, i.e. when statin treatment is started as a first-line care in clinically unstable patients. This therapeutic option is supported by (i) experimental studies, showing a protective effect of statins under the condition of acute ischemia, (ii) analysis of different registers and trials, demonstrating better prognosis of statin-treated patients, and (iii) small clinical trials, describing a lower peri-procedural infarction rate during coronary intervention or lower level of C-reactive protein and other inflammatory markers as a result of statin therapy. Nevertheless, confirmation of this hypothesis by large prospective controlled clinical trials will be necessary before introduction of statins as the first line therapy in unstable patients with ACS, even without knowledge of the blood cholesterol level.

* Address for correspondence: Petr Ostadal, MD, PhD. Dept. of Cardiology, University Hospital Kralovske Vinohrady, Srobarova 50, 100 34, Prague, Czech Republic. e-mail: ostadal@yahoo.com

3-Hydroxy-3-methylglutaryl coenzyme A (HMG-CoA) reductase inhibitors (statins), lower cholesterol level by decreasing the production of low-density lipoproteins (LDL) and up-regulating the expression of the LDL-receptor. Statins are widely used in patients with hyperlipidemia and coronary artery disease (CAD) worldwide. There is strong evidence documenting their efficacy and supporting their use in both primary and secondary prevention for reduction in mortality and nonfatal cardiovascular events. This effect can be, however, only partially explained by the lipid-lowering mechanisms. During the past decade substantial effort has been expended for the investigation of non-lipid-mediated actions of statins, referred to also as "pleiotropic effects". It has been found that beyond the lipid-lowering activity, statins exert anti-inflammatory, antithrombotic, and antioxidant effects, increase nitric oxide (NO) production or improve endothelial dysfunction.

Coronary atherosclerosis has been known for a long time to be associated with lipid metabolism disorder. It has been shown, however, that "classical" risk factors, such as hypercholesterolemia, hypertension, diabetes, and cigarette smoking could not fully explain the development of coronary stenosis in all patients suffering from CAD. Intensive study of the pathogenesis of coronary plaque development and rupture led to the hypothesis of possible involvement of other mechanisms in this process. Particularly inflammation was described as an important factor in the development of CAD and the increased level of some inflammatory markers, e.g. C-reactive protein (CRP), has been shown to be one of the strongest risk factors for cardiovascular events. Activation of inflammatory pathways plays probably also the key role in coronary plaque destabilization and rupture, with subsequent thrombosis development, clinically manifested as ACS. Statins have been shown to modulate several mechanisms, involved in the pathogenesis of ACS, including inflammation or thrombosis (Table 1); the increased interest in the use of statins for the therapy of ACS is therefore fully justified.

Table 1. Possible favorable effects of statins in acute coronary syndrome

- suppression of the production of pro-inflammatory cytokines
- suppression of the expression of adhesive molecules
- inhibition of the production of metalloproteinases
- antioxidant effect
- increased production of NO
- antithrombotic effect

The Vulnerable Plaque

During the past years, marked progress has been made in the understanding of pathogenesis of ACS. Several pathways have been discovered to participate in the development of coronary instability; especially activation of inflammatory mechanisms plays probably an important role in the plaque destabilization. Simultaneously, statins have been shown to influence various factors, participating in this process. The short overview of the current knowledge of the pathogenesis of vulnerable plaque development provides, therefore, the fundamental theoretical background, essential for better understanding of possible therapeutic targets for statins.

Based on the pathological and angiographic studies, plaque fissuring or rupture was determined as a trigger of coronary thrombosis, which is the cause of the majority of ACS. Atherosclerotic plaque susceptible to disruption differs in some histological features: it usually contains a large lipid pool, accumulation of macrophages, and a thin fibrous cap (1-4). The mediators contributing to plaque vulnerability may be divided into extrinsic and intrinsic promoters (5, 6). The extrinsic factors actually trigger plaque disruption; they are represented by circumferential stress, hemodynamic shear stress, vasospasm, plaque fatigue, and prothrombotic milieu (6-9). The intrinsic factors are responsible for susceptibility to plaque rupture or directly cause that event; they include size and composition of lipid core, neovascularization, endothelial erosion/fissure, low fibrous cap thickness, inflammation, matrix degradation enzymes, decreased number of smooth muscle cells (SMC) and low collagen content, outward remodeling, and nodular calcification (6).

Several cellular types are deeply involved in the pathogenesis of vulnerable plaque development.

Macrophages

Lesional macrophages overexpress various enzymes, signaling molecules, pro-inflammatory cytokines, and other active substances. Proteinases derived from macrophages include members of the matrix metalloproteinase (MMP) family, represented by collagenases MMP-1, MMP-8, and MMP-13. Collagenases are supposed to be responsible for fibrous cap degradation (1, 10), associated with the loss of tensile stress tolerance (11). Other enzymes expressed by macrophages are myeloperoxidases, playing a major role in the release of reactive oxygen species, resulting in oxidative stress (OS). OS may activate other cell types such as endothelial cells (EC) and different MMPs (MMP-2, MMP-9, nad MMP-14) (12-14). Macrophages in atheroma secrete also (i) a number of pro-inflammatory cytokines: interleukin 1β (IL-1β), interleukin 2 (IL-2), tumor necrosis factor α (TNF-α), and interferon γ (IFN-γ) and (ii) growth factors: vascular endothelial growth factor (VEGF), macrophage-colony-stimulating factor (M-CSF), leading to activation and/or proliferation of other cell types such as SMC or EC (15). Moreover, lesional macrophages express also the tissue factor (TF), a potent initiator of coagulation cascade, and plasminogen activator inhibitor 1 (PAI-1), inhibitor of fibrinolysis (16, 17). These agents may accelerate thrombus formation after fibrous cap disruption (1).

Endothelium

Another major contributor to the coronary plaque vulnerability are EC. Expression of different surface molecules (vascular cell adhesion molecule 1 (VCAM-1), selectin E, and selectin P) and production of pro-inflammatory substances (monocyte chemoattractant protein 1 (MCP-1)) by EC lead to leukocyte recruitment and formation of macrophage-rich atheroma (18-21). In atheroma is decreased also the expression of endothelial NO synthase (eNOS), producing NO from L-arginine (22).

Smooth Muscle Cells

Proliferation of SMC similarly as production of different components of extracellular matrix by SMC is considered to be important for plaque stability (23). Advanced atherosclerotic lesions are, however, characterized by paucity of SMC proliferation and decreased collagen content (7, 10, 23).

T-lymphocytes

Different types of T-lymphocytes have been detected in atherosclerotic plaques. An important role in the unstable coronary plaque development is probably played by the interactions between various phenotypes of $CD4^+$ T-lymphocytes (Th cells). While Th_1 response, characterized by production of IFN-γ and IL-12, decreases SMC proliferation and matrix synthesis, Th_2 response (production of IL-4, IL-5, and IL-10) inhibits the development of atherosclerosis by possible inhibition of apoptosis and down-regulation of Th_1 response (6, 24, 25).

Pleiotropic Effects of Statins

As mentioned above, statins are highly effective in the treatment of dyslipidemias. They significantly reduce total cholesterol, LDL-cholesterol or triglyceride, and increase HDL-cholesterol levels already after a several-week therapy (26, 27). Statins, however, exert also other effects that seem to be independent of lipid lowering; these effects are referred to as "pleiotropic". During the past several years it has been repeatedly observed that statins could influence various mechanisms that are deeply involved in the vulnerable plaque development. Discovery of the pleiotropic effects of statins fully supports, therefore, the idea to use statins in the therapy of ACS.

Anti-Inflammatory Effects of Statins

As described above, inflammation is probably deeply involved in the pathogenesis of atherosclerosis and in the development of coronary plaque vulnerability. Increased level of a variety of circulating markers of inflammation (CRP, serum amyloid A, heat-shock protein 65, IL-6, and circulating adhesion molecules ICAM-1, V-CAM-1) have been described to be related to the severity of atherosclerosis and to impaired prognosis of patients with ischemic heart disease. The level of CRP, one of the most important risk factors for cardiovascular events, has been repeatedly shown to be reduced by different statins and this reduction was independent of the lipid-lowering effect (28-36). Furthermore, statins suppressed secretion of some cytokines, such as IL-1, IL-6, IL-8, TNF-α or MCP-1 (37-44) as well as production of MMPs (39, 45-47). Moreover, reduction of circulating adhesion molecules ICAM-1, P-selectin, and E-selectin has been described after statin therapy (48-50).

Antioxidant Effects of Statins

Reactive oxygen species are directly involved in the degradation of NO and enhance endothelial dysfunction; moreover, oxidation of LDL contributes to the foam cell formation by increasing the accumulation of cholesterol in macrophages and stimulates thrombosis and inflammation (51). It has been described that fluvastatin exerts a superoxide or hydroxyl radical scavenging activity and reduces susceptibility of LDL to oxidation (52-55), cerivastatin scavenged superoxide and preserved active NO (56, 57); another member of the statin group, atorvastatin, decreased the free radical-induced lipid peroxidation in plasma and increased the total antioxidant status (58).

Statin Effects on Endothelial Dysfunction

Endothelial dysfunction can be recognized as an imbalance between mechanisms, responsible for vasoconstriction and vasodilation. Endothelium-derived vasorelaxation is mediated by NO; in contrast, potent vasoconstrictive agents are endothelin-1 (ET-1) and angiotensin (51).

Statin Effects on Nitric Oxide

NO plays a protective role in CAD by its vasodilating effect, modification of the inflammatory response, decrease of SMC proliferation, and leukocyte and platelet activation (51, 59, 60). Furthermore, NO reduces the endothelial expression of adhesion molecules, monocyte adhesion to the endothelium, and decreases production of IL-6 and IL-8 (51, 61). Increase in NO production, mediated by up-regulation of eNOS, has been observed after administration of statins in different experimental models (56, 57, 62-64); a possible explanation of this effect is the inhibition of G-proteins by statins, leading to reduced eNOS mRNA degradation and higher eNOS level and activity (51, 64-66).

Statins and Endothelin-1

ET-1 is synthesized by endothelial cells and has an action opposite to NO – it stimulates vasoconstriction and vascular cell proliferation and acts as a platelet activator (51, 67). In experimental models, the activation of ET-1 expression promotes the development of atherosclerosis (68). Administration of statins induces a down-regulation of the pre-pro ET-1-mRNA levels (69); it is not clear, however, whether this is a direct effect of statins on ET-1 synthesis or whether it is mediated by statin-induced increase of NO production that exerts a similar effect on ET-1 (51, 69).

In clinical settings, the effect of statins on endothelial dysfunction has been assessed mostly by the measurement of flow-mediated dilatation (FMD); this parameter is impaired in patients with atherosclerosis (70, 71). Administration of statins in patients with hypercholesterolemia and/or CAD significantly improved the FMD; this effect is probably independent of statin-induced changes in lipid levels (70-74).

Antithrombotic Effects of Statins

It has been repeatedly shown in cultured monocytes/macrophages, SMC or endothelial cells that statins can reduce the level and activity of TF, which initiates the coagulation cascade by activating factors IX and X (46, 75-79). This effect of statins was confirmed also *in vivo* on animal models and in human studies (46, 79-83). On the other hand, statins decrease the level of the total tissue factor pathway inhibitor (TFPI), a potent anticoagulant agent (84-87); they do not, however, influence the free TFPI (84, 86, 87). Because the anticoagulant activity of TFPI is related just to the free fraction of TFPI (87, 88), alteration of the total TFPI seems to play a relatively marginal role in the global anticoagulant activity of statins (88).

Based on the studies detecting the effect of statins on prothrombin fragment (F1+2), fibrinopeptide A (FPA), and thrombin-antithrombin III (TAT) complexes, statins significantly decrease thrombin formation (82, 89-94). It has been shown that this effect results not only from the inhibition of platelet-dependent thrombin formation but also from the decreased TF expression (76, 88).

In the clinical trials, however, there are contradictory opinions on the effect of statins on fibrinogen levels, one of the important risk factors of cardiovascular events (95-99). This discrepancy could be at least partially explained by different methods used in these studies (88). Also factor VII, factor VIII, and von Willebrand factor (vWF) have been identified as predictors of increased risk of CAD (100). Although the literature data dealing with the effect of statins on these factors are inconsistent, the majority of studies have shown favorable results (88).

Statins, however, influence not only the coagulation but also the fibrinolytic activity. It has been repeatedly described by *in vitro* studies that administration of statins results in the increased level, activity, and synthesis of tissue plasminogen activator (tPA), and simultaneously in the decreased level, activity and synthesis of PAI-1 (63, 99, 101-107). Although there is a general agreement among the *in vitro* studies concerning the fibrinolysis-stimulating role of statins, clinical trials have produced less convincing results (88). Clinical studies are, however, inconsistent in inclusion criteria, type of statin and its dosage, duration of treatment or in the methodological approaches (88).

Statins in Early Secondary Prevention

There is overwhelming evidence about the beneficial role of statins in the therapy of stable forms of CAD, arising from the large prospective trials as 4S (Scandinavian Simvastatin Survival Study) (108), CARE (Cholesterol and Recurrent Events) (109) or LIPID (Long-Term Intervention with Pravastatin in Ischaemic Disease) (110). The discovery of the non-lipid effects of statins in the recent years was, however, immediately reflected in the efforts to use statins also in less stabilized patients. Different observational studies have been published, showing the favorable effect of statin therapy started soon after ACS; furthermore, several randomized prospective trials have been carried out to evaluate the effect of statins, when administered early after ACS.

Observational Studies

A large, prospective, cohort study using data from the Swedish Register of Cardiac Intensive Care (RIKS-HIA) demonstrated that early initiation of statin therapy in patients with acute myocardial infarction was associated with a reduced 1-year mortality (RR 0.75, $p=0.001$) (111). Post-hoc analysis of two randomized trials – GUSTO IIb (Global Use of Strategies to Open Occluded Coronary Arteries) and PURSUIT (Platelet Glycoprotein IIb/IIIa in Unstable Angina: Receptor Suppression Using Integrilin Therapy) – showed that patients with ACS discharged on lipid-lowering drugs have a survival benefit at 6 months (RR 0.48, $p<0.001$) (112). In the OPUS/TIMI 16 study (Orofiban in Patients with Unstable Coronary Syndromes/Thrombolysis in Myocardial Infarction), mortality at 1 month was reduced in patients treated with lipid-lowering drugs (RR 0.30, $p<0.001$) (113). The analysis of SYMPHONY trial (Sibrafiban Versus Aspirin to Yield Maximum Protection from Ischemic Heart Events Post-Acute Coronary Syndromes) showed lower mortality in statin-treated patients at 3 months (RR 0.58, $p<0.05$) (114). Subgroup analysis of the PRISM study (Platelet Receptor Inhibition in Ischemic Syndrome Management) demonstrated better prognosis of statin-treated patients compared with patients without statin therapy; surprisingly, the worst prognosis in this analysis was described in patients in whom statin therapy was withdrawn after hospital admission for ACS (115). Early use of statin in acute myocardial infarction was associated with a reduced in-hospital mortality in MITRA registry (Maximized Individual Therapies in Acute Myocardial Ischemia) (116). It has been shown that statin therapy also improved prognosis in patients undergoing percutaneous coronary intervention (117). Moreover, analysis of the Euro Heart Survey on acute coronary syndrome showed reduced 7-day mortality in patients with ACS receiving statin therapy within 24 hours of admission (118).

Randomized Trials

Recently, several randomized trials, dealing with the statin therapy, started early after ACS, have been published. In a short survey we would like to summarize their design and results of primary endpoints (Table 2).

The Pravastatin Turkish Trial (PTT) (119) was a small study that randomized 150 patients with acute myocardial infarction (AMI) to pravastatin 40 mg or no lipid-lowering drug within 6 hours of hospital admission. All patients were treated with intravenous fibrinolytic therapy. At 6 months no difference was found in mortality; the pravastatin-treated group experienced, however, less recurrent ischemic events compared with the placebo group.

Table 2. Large randomized trials concerning with the early statin therapy after acute coronary syndrome

Trial	No. of pts.	Drugs studied	Primary endpoint	Results
MIRACL	3086	atorvastatin/placebo	MACE	RR 0.84, p=0.048
FLORIDA	540	fluvastatin/placebo	ischemia	n.s.
PROVE-IT	4162	atorvastatin/pravastatin	MACE	RR 0.84, p=0.005
PACT	3408	pravastatin/placebo	MACE	n.s.
A to Z	4497	simvastatin/placebo	MACE	n.s.
PRINCESS	3600	cerivastatin/placebo	MACE	n.s.

MIRACL, the Myocardial Ischemia Reduction with Aggressive Cholesterol Lowering trial; FLORIDA, the Fluvastatin on Risk Diminishing After Acute Myocardial Infarction trial; PROVE-IT, the Pravastatin or Atorvastatin Evaluation and Infection Therapy-Thrombolysis in Myocardial Infarction 22 trial; PACT, the Pravastatin in Acute Coronary Treatment trial; PRINCESS, the Prevention of Ischemic Events by Early Treatment of Cerivastatin Study; MACE, major adverse cardiac events.

The Myocardial Ischemia Reduction with Aggressive Cholesterol Lowering (MIRACL) trial (120) randomized 3086 patients with unstable angina or non-Q-wave AMI to atorvastatin 80 mg or placebo within 24 to 96 hours after hospital admission; treatment continued for 16 weeks. The investigators showed a 16 % relative reduction in the combined endpoint of death, recurrent nonfatal AMI, resuscitated cardiac arrest, or recurrent ischemia requiring repeated hospitalization (14.8 % vs. 17.4 %; RR 0.84; p=0.048). This benefit was largely driven by a reduction in the endpoint of recurrent ischemia requiring hospitalization (6.2 % vs. 8.4 %; RR 0.74; p=0.02). Limitation of this study from the current point of view was the exclusion criterion of coronary intervention undergone or planned; only conservatively treated patients were included in this study.

The Fluvastatin on Risk Diminishing After Acute Myocardial Infarction (FLORIDA) trial (121) randomized 540 patients with AMI to fluvastatin or placebo within 14 days of the index MI. The primary endpoint was reduction in cardiac ischemia, determined by ambulatory Holter monitoring. No differences in occurrence of ischemia were observed and, similarly, there were no differences in mortality and major adverse cardiac events (MACE) rate after 1 year.

The Lipid-Coronary Artery Disease (L-CAD) trial (122) was primarily designed to examine the effect of pravastatin on angiographic regression of atherosclerosis. In the L-CAD study, 126 patients, who had undergone percutaneous coronary intervention (PCI) for ACS, were randomized to pravastatin 20-40 mg (with or without cholestyramine and/or nicotinic acid) or to usual care determined by family physician; patients were enrolled by hospital discharge. After 2 years of follow-up, fewer patients in the pravastatin group experienced MACE as compared with the usual care group (OR 0.28, p=0.005).

The Lescol Intervention Prevention Study (LIPS) (123) enrolled 1677 patients, scheduled for a first PCI procedure for stable angina, unstable angina or silent ischemia. Patients were randomized to fluvastatin 40 mg twice daily or placebo at an average of 2.7 days after the procedure. Statin therapy resulted in a reduction of the occurrence of MACE at a median follow-up time of 3.9 years (21.4% vs. 26.7%; RR 0.78; p=0.01).

The Pravastatin or Atorvastatin Evaluation and Infection Therapy-Thrombolysis in Myocardial Infarction 22 (PROVE IT-TIMI 22) trial (124) randomized 4162 patients within 10 days of hospital admission for ACS to pravastatin 40 mg daily (standard therapy) or atorvastatin 80 mg daily (intensive therapy). After a median follow-up of 2 years, the rate of MACE was significantly lower in the intensive therapy group as compared with the standard therapy (22.4% vs. 26.3%; RR 0.84; p=0.005).

The Pravastatin in Acute Coronary Treatment (PACT) trial (125) tested the effect of pravastatin administration within 24 hours of the onset of symptoms in patients with ACS. The recruitment of 10,000 patients was planned; the study was, however, stopped early. A total of 3408 patients were randomized to pravastatin 20-40 mg or placebo. After 30 days of follow-up, no significant difference in the occurrence of MACE was observed.

Phase Z of the A to Z trial (126) compared an early intensive vs. a delayed conservative strategy of simvastatin therapy in patients with ACS. Within 5 days of the onset of symptoms, a total of 4497 patients were randomized to simvastatin 40 mg daily for 1 month followed by 80 mg daily thereafter or to placebo for 4 months followed by simvastatin 20 mg daily. The rates of MACE were not significantly different between the early intensive and the delayed conservative group either at 4 months or at 2 years.

In the Prevention of Ischemic Events by Early Treatment of Cerivastatin Study (PRINCESS) (127) patients were randomized within 48 hours of admission for ACS to cerivastatin or placebo. This trial was prematurely interrupted after cerivastatin was withdrawn from worldwide market. Data obtained from a total of 3,600 patients followed up for 4.5 months were analyzed; no significant difference in the occurrence of MACE was found between the groups.

It can be concluded that these prospective trials have shown safety and in some points also benefit of the early initiation of statin therapy after ACS (Table 2). It is, however, necessary to stress that these trials were designed for early (in some cases "very early") secondary prevention, not for the real therapy of ACS. In all of the above-mentioned studies, patients were randomized at least several hours and in most cases even several days after hospital admission (Figure 1). During such a long period of time the majority of patients are already clinically stabilized. In the context of the above described pathogenesis of vulnerable plaque development and the mechanisms of pleiotropic effects of statins it is obvious that there are relevant reasons to use these drugs not only in the secondary prevention but also directly in the therapy of ACS, when administered to really unstable patients.

STATINS IN THE THERAPY OF ACS

As discussed above, pleiotropic effects of statins may affect different pathogenic pathways, participating in the development of vulnerable plaque and in the pathogenesis of ACS; expanding this suggestion, statins may significantly contribute to plaque stabilization, reduction of thrombus formation and acceleration of fibrinolysis. It is, therefore, fully justified to investigate the possible favorable effect of statin therapy at the time of coronary instability and to evaluate the shift of statin administration from hospital discharge to the first-line therapy, beside aspirin (128).

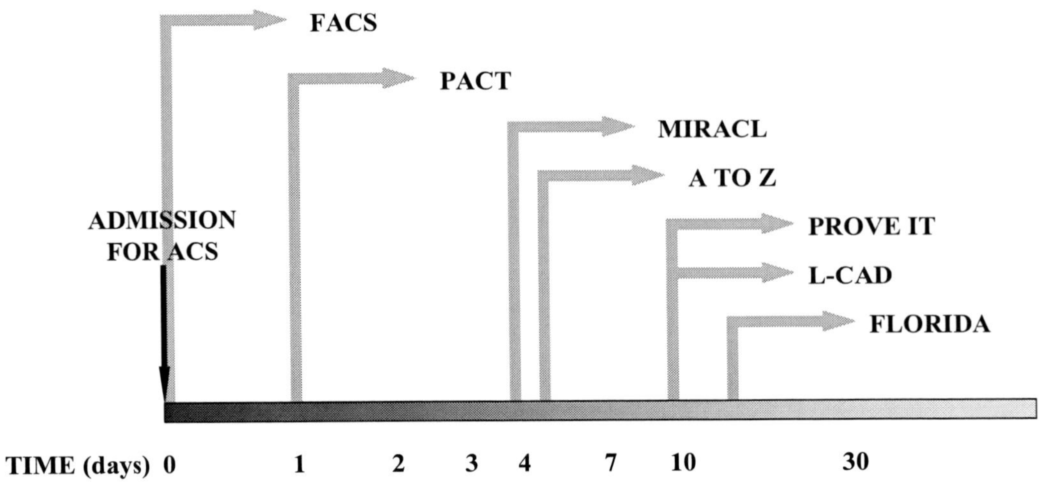

Figure 1. Relation of the randomization time to the time of admission in randomized trials with statins and ACS. In the Fluvastatin in the therapy of Acute Coronary Syndrome (FACS) trial patients are randomized at admission; in Pravastatin in Acute Coronary Treatment (PACT) trial statin therapy was initiated within 24 hours of onset of ACS; in Myocardial Ischemia Reduction With Aggressive Cholesterol Lowering (MIRACL) trial patients were randomized 24 to 96 hours after ACS; in Z-phase of the A-to-Z trial simvastatin therapy was initiated within 5 days of the onset of ACS, after clinical stabilization; in Pravastatin or Atorvastatin Evaluation and Infection Therapy (PROVE IT) trial patients were randomized up to 10 days after ACS; in Lipid-Coronary Artery Disease (L-CAD) trial statin therapy was initiated up to ten days from ACS; in Fluvastatin on Risk Diminishing After Acute Myocardial Infarction (FLORIDA) trial patients were randomized up to two weeks after ACS.

Animal Studies

It has been repeatedly shown that treatment of animals with statins prior to the onset of myocardial ischemia reduces ischemia-reperfusion injury. Already in 1999, Lefer et al. (129) have shown in the model of isolated perfused rat heart a cardioprotective effect of simvastatin, when administered before the induction of ischemia. Simvastatin inhibited leukocyte-endothelium interactions and improved contractile parameters. Ueda et al. (130) described that pravastatin treatment decreased the infarct size after experimentally induced ischemia and ischemic preconditioning in hypercholesterolemic rabbits. In another paper Lefer et al. (131) showed reduction in the infarct size after simvastatin treatment in the model of diabetic mouse. It has been also found that this protective effect may be at least partly mediated by the stimulation of eNOS and increased production of NO (131, 132).

While the above experimental studies demonstrate the impact of prophylactic therapy they do not address the question, whether patients with AMI might benefit from the initiation of statin therapy after the onset of ischemia or prior to reperfusion. Bauersachs et al. (133) failed to demonstrate a reduction of infarct size when the statin was administered 24 hours after onset of myocardial ischemia, similarly to the protocol of most of the randomized clinical trials. Probably the first study, showing a significant benefit of the statin therapy, started immediately after the onset of ischemia was published by Hayashidani et al. (134)

They described lower mortality in mice subjected to coronary artery ligation and fluvastatin therapy, introduced just after the procedure. Furthermore, fluvastatin in this study attenuated left ventricular remodelation, decreased the incidence of heart failure, and reduced activity of MMPs. Bell and Yellon (135) demonstrated the beneficial effect of atorvastatin, administered at the time of reperfusion: in the experimental model of isolated perfused mouse heart atorvastatin reduced ischemia-reperfusion injury (infarct size), when added to the perfusion solution. Results of this *in vitro* study were recently confirmed by Wolfrum et al. (136) in their *in vivo* study of experimental infarction in rats. They have administered activated simvastatin intravenously 3 min before restoration of flow after a temporary coronary artery occlusion; simvastatin treatment decreased infarct size by 42 %. The two latest studies have shown also another important aspect of the immediate effect of statin therapy: they have demonstrated an acute increase of eNOS activity, which is probably not related to the up-regulation of eNOS by stabilizing eNOS mRNA. To stabilize eNOS mRNA, several hours of statin treatment are essential as was shown in a cell culture system (62). The acute stimulation of eNOS activity is, however, at least partially mediated by PI 3-kinase/Akt pathway which is capable to phosphorylate eNOS within minutes (135, 136).

Clinical Studies

No literary data on the administration of statins as a part of the first-line therapy in patients with ACS were available. We arranged, therefore, a pilot, small, prospective trial in 44 consecutive patients with ACS without ST-elevation, randomized at admission to cerivastatin 0.3 mg or no statin therapy (137). Already 24 hours after hospital admission (and after the initiation of statin therapy), a significant reduction in the level of CRP and IL-6 was observed in the cerivastatin-treated group as compared with the untreated group. On the other hand, the level of IL-8 was not influenced by cerivastatin in this setting. A strong effect of the low dose cerivastatin therapy for only 24 hours was the main reason for organizing the Fluvastatin in the Therapy of Acute Coronary Syndrome (FACS) trial (138). This is an ongoing multicenter, randomized, double-blind, placebo-controlled study comparing the effect of 30-day 80 mg fluvastatin therapy starting at the time of hospital admission with placebo. The primary endpoint is the influence of therapy on inflammatory markers (CRP and interleukin-6) and on pregnancy-associated plasma protein A (PAPP-A), a combined secondary endpoint is 30-day and one-year occurrence of death, nonfatal myocardial infarction, recurrent symptomatic ischemia, urgent revascularization, and cardiac arrest.

Conclusion

Statins were introduced into the clinical practice as lipid-lowering drugs for treatment of high blood cholesterol. They were shown to be highly effective in hypercholesterolemic patients for primary and secondary prevention of CAD. Their efficacy in secondary prevention was demonstrated in the large prospective clinical trials in stable CAD; these studies enrolled patients at least several months after an ACS. Later, it has been observed that statins exert a favorable effect not only in hypercholesterolemia, but also in patients with normal or low cholesterol and, therefore, possible non-lipid effects of statins have been suggested. The discovery of pleiotropic effects of statins opened the new field of indication for statin treatment. Recently, several randomized trials have been published, showing safety and in some points also efficacy of statin therapy started early after ACS. Extension of knowledge of the pleiotropic effects of statins together with the increasing understanding of pathogenesis of ACS have been shifting, however, the initiation of statin therapy closer to the onset of symptoms. Recent experimental studies, similarly as the first clinical trials have brought promising results, supporting the idea of cardioprotective effect of statin administration in the first-line therapy of ACS. Confirmation of this approach by a large randomized trial is, however, needed; nevertheless, based on the currently available data, statins have a high chance of achieving a similar place in the therapy of ACS as the pillar of contemporary therapeutic strategy, aspirin.

Acknowledgements

This work was supported by the Charles University in Prague Research Project MSM 0021620817 awarded by the Ministry of Education, Youth and Physical Education of the Czech Republic and by a grant from the Ministry of Health of the Czech Republic, No. 00064203.

References

[1] Aikawa M, Libby P. The vulnerable atherosclerotic plaque: pathogenesis and therapeutic approach. *Cardiovasc Pathol* 2004;13(3):125-38.

[2] Davies MJ. Acute coronary thrombosis--the role of plaque disruption and its initiation and prevention. *Eur Heart J* 1995;16 Suppl L:3-7.

[3] Davies MJ. Detecting vulnerable coronary plaques. *Lancet* 1996;347(9013):1422-3.

[4] Ambrose JA, Hjemdahl-Monsen CE, Borrico S, Gorlin R, Fuster V. Angiographic demonstration of a common link between unstable angina pectoris and non-Q-wave acute myocardial infarction. *Am J Cardiol* 1988;61(4):244-7.

[5] Pasterkamp G, Schoneveld AH, Hijnen DJ, de Kleijn DP, Teepen H, van der Wal AC, et al. Atherosclerotic arterial remodeling and the localization of macrophages and matrix metalloproteases 1, 2 and 9 in the human coronary artery. *Atherosclerosis* 2000;150(2):245-53.

[6] Dickson BC, Gotlieb AI. Towards understanding acute destabilization of vulnerable atherosclerotic plaques. *Cardiovasc Pathol* 2003;12(5):237-48.

[7] Falk E, Shah PK, Fuster V. Coronary plaque disruption. Circulation 1995;92(3):657-71.

[8] Rabbani R, Topol EJ. Strategies to achieve coronary arterial plaque stabilization. *Cardiovasc Res* 1999;41(2):402-17.

[9] Bank AJ, Versluis A, Dodge SM, Douglas WH. Atherosclerotic plaque rupture: a fatigue process? *Med Hypotheses* 2000;55(6):480-4.

[10] Shah PK, Falk E, Badimon JJ, Fernandez-Ortiz A, Mailhac A, Villareal-Levy G, et al. Human monocyte-derived macrophages induce collagen breakdown in fibrous caps of atherosclerotic plaques. Potential role of matrix-degrading metalloproteinases and implications for plaque rupture. *Circulation* 1995;92(6):1565-9.

[11] Loree HM, Kamm RD, Stringfellow RG, Lee RT. Effects of fibrous cap thickness on peak circumferential stress in model atherosclerotic vessels. *Circ Res* 1992;71(4):850-8.

[12] Heinecke JW. Oxidized amino acids: culprits in human atherosclerosis and indicators of oxidative stress. *Free Radic Biol Med* 2002;32(11):1090-101.

[13] Sugiyama S, Okada Y, Sukhova GK, Virmani R, Heinecke JW, Libby P. Macrophage myeloperoxidase regulation by granulocyte macrophage colony-stimulating factor in human atherosclerosis and implications in acute coronary syndromes. *Am J Pathol* 2001;158(3):879-91.

[14] Rajagopalan S, Meng XP, Ramasamy S, Harrison DG, Galis ZS. Reactive oxygen species produced by macrophage-derived foam cells regulate the activity of vascular matrix metalloproteinases in vitro. Implications for atherosclerotic plaque stability. *J Clin Invest* 1996;98(11):2572-9.

[15] Libby P. Atherosclerosis: the new view. *Sci Am* 2002;286(5):46-55.

[16] Wilcox JN, Smith KM, Schwartz SM, Gordon D. Localization of tissue factor in the normal vessel wall and in the atherosclerotic plaque. *Proc Natl Acad Sci USA* 1989;86(8):2839-43.

[17] Lupu F, Bergonzelli GE, Heim DA, Cousin E, Genton CY, Bachmann F, et al. Localization and production of plasminogen activator inhibitor-1 in human healthy and atherosclerotic arteries. *Arterioscler Thromb* 1993;13(7):1090-100.

[18] Cybulsky MI, Iiyama K, Li H, Zhu S, Chen M, Iiyama M, et al. A major role for VCAM-1, but not ICAM-1, in early atherosclerosis. *J Clin Invest* 2001;107(10):1255-62.

[19] Gu L, Okada Y, Clinton SK, Gerard C, Sukhova GK, Libby P, et al. Absence of monocyte chemoattractant protein-1 reduces atherosclerosis in low density lipoprotein receptor-deficient mice. *Mol Cell* 1998;2(2):275-81.

[20] Plutzky J. Inflammatory pathways in atherosclerosis and acute coronary syndromes. *Am J Cardiol* 2001;88(8A):10K-15K.

[21] Kadar A, Glasz T. Development of atherosclerosis and plaque biology. *Cardiovasc Surg* 2001;9(2):109-21.

[22] Oemar BS, Tschudi MR, Godoy N, Brovkovich V, Malinski T, Luscher TF. Reduced endothelial nitric oxide synthase expression and production in human atherosclerosis. *Circulation* 1998;97(25):2494-8.

[23] Braganza DM, Bennett MR. New insights into atherosclerotic plaque rupture. *Postgrad Med J* 2001;77(904):94-8.

[24] Libby P. Changing concepts of atherogenesis. *J Intern Med* 2000;247(3):349-58.

[25] Pinderski LJ, Fischbein MP, Subbanagounder G, Fishbein MC, Kubo N, Cheroutre H, et al. Overexpression of interleukin-10 by activated T lymphocytes inhibits atherosclerosis in LDL receptor-deficient Mice by altering lymphocyte and macrophage phenotypes. *Circ Res* 2002;90(10):1064-71.

[26] Jones P, Kafonek S, Laurora I, Hunninghake D. Comparative dose efficacy study of atorvastatin versus simvastatin, pravastatin, lovastatin, and fluvastatin in patients with hypercholesterolemia (the CURVES study). *Am J Cardiol* 1998;81(5):582-7.

[27] Insull W, Kafonek S, Goldner D, Zieve F. Comparison of efficacy and safety of atorvastatin (10mg) with simvastatin (10mg) at six weeks. ASSET Investigators. *Am J Cardiol* 2001;87(5):554-9.

[28] Albert MA, Danielson E, Rifai N, Ridker PM. Effect of statin therapy on C-reactive protein levels: the pravastatin inflammation/CRP evaluation (PRINCE): a randomized trial and cohort study. *Jama* 2001;286(1):64-70.

[29] Kinlay S, Schwartz GG, Olsson AG, Rifai N, Leslie SJ, Sasiela WJ, et al. High-dose atorvastatin enhances the decline in inflammatory markers in patients with acute coronary syndromes in the MIRACL study. *Circulation* 2003;108(13):1560-6.

[30] Ridker PM, Rifai N, Lowenthal SP. Rapid reduction in C-reactive protein with cerivastatin among 785 patients with primary hypercholesterolemia. *Circulation* 2001;103(9):1191-3.

[31] Jialal I, Stein D, Balis D, Grundy SM, Adams-Huet B, Devaraj S. Effect of hydroxymethyl glutaryl coenzyme a reductase inhibitor therapy on high sensitive C-reactive protein levels. *Circulation* 2001;103(15):1933-5.

[32] Riesen WF, Engler H, Risch M, Korte W, Noseda G. Short-term effects of atorvastatin on C-reactive protein. *Eur Heart J* 2002;23(10):794-9.

[33] van de Ree MA, Huisman MV, Princen HM, Meinders AE, Kluft C. Strong decrease of high sensitivity C-reactive protein with high-dose atorvastatin in patients with type 2 diabetes mellitus. *Atherosclerosis* 2003;166(1):129-35.

[34] Koh KK, Schenke WH, Waclawiw MA, Csako G, Cannon RO, 3rd. Statin attenuates increase in C-reactive protein during estrogen replacement therapy in postmenopausal women. *Circulation* 2002;105(13):1531-3.

[35] Kent SM, Flaherty PJ, Coyle LC, Markwood TT, Taylor AJ. Effect of atorvastatin and pravastatin on serum C-reactive protein. *Am Heart J* 2003;145(2):e8.

[36] Ansell BJ, Watson KE, Weiss RE, Fonarow GC. hsCRP and HDL effects of statins trial (CHEST): rapid effect of statin therapy on C-reactive protein and high-density lipoprotein levels A clinical investigation. *Heart Dis* 2003;5(1):2-7.

[37] Romano M, Diomede L, Sironi M, Massimiliano L, Sottocorno M, Polentarutti N, et al. Inhibition of monocyte chemotactic protein-1 synthesis by statins. *Lab Invest* 2000;80(7):1095-100.

[38] Blake GJ, Ridker PM. Novel clinical markers of vascular wall inflammation. *Circ Res* 2001;89(9):763-71.

[39] Koh KK, Son JW, Ahn JY, Choi YM, Jin DK, Park GS, et al. Non-lipid effects of statin on hypercholesterolemic patients established to have coronary artery disease who remained hypercholesterolemic while eating a step-II diet. *Coron Artery Dis* 2001;12(4):305-11.

[40] Rezaie-Majd A, Maca T, Bucek RA, Valent P, Muller MR, Husslein P, et al. Simvastatin reduces expression of cytokines interleukin-6, interleukin-8, and monocyte chemoattractant protein-1 in circulating monocytes from hypercholesterolemic patients. *Arterioscler Thromb Vasc Biol* 2002;22(7):1194-9.

[41] Brull DJ, Sanders J, Rumley A, Lowe GD, Humphries SE, Montgomery HE. Statin therapy and the acute inflammatory response after coronary artery bypass grafting. *Am J Cardiol* 2001;88(4):431-3.

[42] Ikeda U, Shimada K. Statins and monocytes. Lancet 1999;353(9169):2070.

[43] Ferro D, Parrotto S, Basili S, Alessandri C, Violi F. Simvastatin inhibits the monocyte expression of proinflammatory cytokines in patients with hypercholesterolemia. *J Am Coll Cardiol* 2000;36(2):427-31.

[44] Bustos C, Hernandez-Presa MA, Ortego M, Tunon J, Ortega L, Perez F, et al. HMG-CoA reductase inhibition by atorvastatin reduces neointimal inflammation in a rabbit model of atherosclerosis. *J Am Coll Cardiol* 1998;32(7):2057-64.
[45] Ikeda U, Shimpo M, Ohki R, Inaba H, Takahashi M, Yamamoto K, et al. Fluvastatin inhibits matrix metalloproteinase-1 expression in human vascular endothelial cells. *Hypertension* 2000;36(3):325-9.
[46] Aikawa M, Rabkin E, Sugiyama S, Voglic SJ, Fukumoto Y, Furukawa Y, et al. An HMG-CoA reductase inhibitor, cerivastatin, suppresses growth of macrophages expressing matrix metalloproteinases and tissue factor in vivo and in vitro. *Circulation* 2001;103(2):276-83.
[47] Crisby M, Nordin-Fredriksson G, Shah PK, Yano J, Zhu J, Nilsson J. Pravastatin treatment increases collagen content and decreases lipid content, inflammation, metalloproteinases, and cell death in human carotid plaques: implications for plaque stabilization. *Circulation* 2001;103(7):926-33.
[48] Seljeflot I, Tonstad S, Hjermann I, Arnesen H. Reduced expression of endothelial cell markers after 1 year treatment with simvastatin and atorvastatin in patients with coronary heart disease. *Atherosclerosis* 2002;162(1):179-85.
[49] Sardo MA, Castaldo M, Cinquegrani M, Bonaiuto M, Maesano A, Versace A, et al. Effects of atorvastatin treatment on sICAM-1 and plasma nitric oxide levels in hypercholesterolemic subjects. *Clin Appl Thromb Hemost* 2002;8(3):257-63.
[50] Niwa S, Totsuka T, Hayashi S. Inhibitory effect of fluvastatin, an HMG-CoA reductase inhibitor, on the expression of adhesion molecules on human monocyte cell line. *Int J Immunopharmacol* 1996;18(11):669-75.
[51] Tsiara S, Elisaf M, Mikhailidis DP. Early vascular benefits of statin therapy. *Curr Med Res Opin* 2003;19(6):540-56.
[52] Suzumura K, Yasuhara M, Narita H. Superoxide anion scavenging properties of fluvastatin and its metabolites. *Chem Pharm Bull* (Tokyo) 1999;47(10):1477-80.
[53] Suzumura K, Yasuhara M, Tanaka K, Odawara A, Narita H, Suzuki T. An in vitro study of the hydroxyl radical scavenging property of fluvastatin, and HMG-CoA reductase inhibitor. *Chem Pharm Bull* (Tokyo) 1999;47(7):1010-2.
[54] Suzumura K, Yasuhara M, Tanaka K, Suzuki T. Protective effect of fluvastatin sodium (XU-62-320), a 3-hydroxy-3-methylglutaryl coenzyme A (HMG-CoA) reductase inhibitor, on oxidative modification of human low-density lipoprotein in vitro. *Biochem Pharmacol* 1999;57(6):697-703.
[55] Hussein O, Schlezinger S, Rosenblat M, Keidar S, Aviram M. Reduced susceptibility of low density lipoprotein (LDL) to lipid peroxidation after fluvastatin therapy is associated with the hypocholesterolemic effect of the drug and its binding to the LDL. *Atherosclerosis* 1997;128(1):11-8.
[56] Kalinowski L, Dobrucki IT, Malinski T. Cerivastatin potentiates nitric oxide release and enos expression through inhibition of isoprenoids synthesis. *J Physiol Pharmacol* 2002;53(4 Pt 1):585-95.
[57] Kalinowski L, Dobrucki LW, Brovkovych V, Malinski T. Increased nitric oxide bioavailability in endothelial cells contributes to the pleiotropic effect of cerivastatin. *Circulation* 2002;105(8):933-8.
[58] Fuhrman B, Koren L, Volkova N, Keidar S, Hayek T, Aviram M. Atorvastatin therapy in hypercholesterolemic patients suppresses cellular uptake of oxidized-LDL by differentiating monocytes. *Atherosclerosis* 2002;164(1):179-85.

[59] Vaughan CJ, Murphy MB, Buckley BM. Statins do more than just lower cholesterol. *Lancet* 1996;348(9034):1079-82.

[60] Jeremy JY, Rowe D, Emsley AM, Newby AC. Nitric oxide and the proliferation of vascular smooth muscle cells. *Cardiovasc Res* 1999;43(3):580-94.

[61] De Caterina R, Libby P, Peng HB, Thannickal VJ, Rajavashisth TB, Gimbrone MA, Jr., et al. Nitric oxide decreases cytokine-induced endothelial activation. Nitric oxide selectively reduces endothelial expression of adhesion molecules and proinflammatory cytokines. *J Clin Invest* 1995;96(1):60-8.

[62] Laufs U, La Fata V, Plutzky J, Liao JK. Upregulation of endothelial nitric oxide synthase by HMG CoA reductase inhibitors. *Circulation* 1998;97(12):1129-35.

[63] Morikawa S, Takabe W, Mataki C, Kanke T, Itoh T, Wada Y, et al. The effect of statins on mRNA levels of genes related to inflammation, coagulation, and vascular constriction in HUVEC. Human umbilical vein endothelial cells. *J Atheroscler Thromb* 2002;9(4):178-83.

[64] Laufs U, Gertz K, Dirnagl U, Bohm M, Nickenig G, Endres M. Rosuvastatin, a new HMG-CoA reductase inhibitor, upregulates endothelial nitric oxide synthase and protects from ischemic stroke in mice. *Brain Res* 2002;942(1-2):23-30.

[65] Laufs U, Endres M, Custodis F, Gertz K, Nickenig G, Liao JK, et al. Suppression of endothelial nitric oxide production after withdrawal of statin treatment is mediated by negative feedback regulation of rho GTPase gene transcription. *Circulation* 2000;102(25):3104-10.

[66] Laufs U, Liao JK. Targeting Rho in cardiovascular disease. Circ Res 2000;87(7):526-8.

[67] Jagroop IA, Mikhailidis DP. Effect of endothelin-1 on human platelet shape change: reversal of activation by naftidrofuryl. *Platelets* 2000;11(5):272-7.

[68] Barton M, Lattmann T, d'Uscio LV, Luscher TF, Shaw S. Inverse regulation of endothelin-1 and nitric oxide metabolites in tissue with aging: implications for the age-dependent increase of cardiorenal disease. *J Cardiovasc Pharmacol* 2000;36(5 Suppl 1):S153-6.

[69] Hernandez-Perera O, Perez-Sala D, Navarro-Antolin J, Sanchez-Pascuala R, Hernandez G, Diaz C, et al. Effects of the 3-hydroxy-3-methylglutaryl-CoA reductase inhibitors, atorvastatin and simvastatin, on the expression of endothelin-1 and endothelial nitric oxide synthase in vascular endothelial cells. *J Clin Invest* 1998;101(12):2711-9.

[70] Dupuis J, Tardif JC, Cernacek P, Theroux P. Cholesterol reduction rapidly improves endothelial function after acute coronary syndromes. The RECIFE (reduction of cholesterol in ischemia and function of the endothelium) trial. *Circulation* 1999;99(25):3227-33.

[71] de Jongh S, Lilien MR, op't Roodt J, Stroes ES, Bakker HD, Kastelein JJ. Early statin therapy restores endothelial function in children with familial hypercholesterolemia. *J Am Coll Cardiol* 2002;40(12):2117-21.

[72] Marchesi S, Lupattelli G, Siepi D, Schillaci G, Vaudo G, Roscini AR, et al. Short-term atorvastatin treatment improves endothelial function in hypercholesterolemic women. *J Cardiovasc Pharmacol* 2000;36(5):617-21.

[73] Tsunekawa T, Hayashi T, Kano H, Sumi D, Matsui-Hirai H, Thakur NK, et al. Cerivastatin, a hydroxymethylglutaryl coenzyme a reductase inhibitor, improves endothelial function in elderly diabetic patients within 3 days. *Circulation* 2001;104(4):376-9.

[74] Omori H, Nagashima H, Tsurumi Y, Takagi A, Ishizuka N, Hagiwara N, et al. Direct in vivo evidence of a vascular statin: a single dose of cerivastatin rapidly increases

vascular endothelial responsiveness in healthy normocholesterolaemic subjects. *Br J Clin Pharmacol* 2002;54(4):395-9.

[75] Colli S, Eligini S, Lalli M, Camera M, Paoletti R, Tremoli E. Vastatins inhibit tissue factor in cultured human macrophages. A novel mechanism of protection against atherothrombosis. *Arterioscler Thromb Vasc Biol* 1997;17(2):265-72.

[76] Ferro D, Basili S, Alessandri C, Cara D, Violi F. Inhibition of tissue-factor-mediated thrombin generation by simvastatin. *Atherosclerosis* 2000;149(1):111-6.

[77] Nagata K, Ishibashi T, Sakamoto T, Ohkawara H, Shindo J, Yokoyama K, et al. Rho/Rho-kinase is involved in the synthesis of tissue factor in human monocytes. *Atherosclerosis* 2002;163(1):39-47.

[78] Eto M, Kozai T, Cosentino F, Joch H, Luscher TF. Statin prevents tissue factor expression in human endothelial cells: role of Rho/Rho-kinase and Akt pathways. *Circulation* 2002;105(15):1756-9.

[79] Camera M, Toschi V, Comparato C, Baetta R, Rossi F, Fuortes M, et al. Cholesterol-induced thrombogenicity of the vessel wall: inhibitory effect of fluvastatin. *Thromb Haemost* 2002;87(4):748-55.

[80] Baetta R, Camera M, Comparato C, Altana C, Ezekowitz MD, Tremoli E. Fluvastatin reduces tissue factor expression and macrophage accumulation in carotid lesions of cholesterol-fed rabbits in the absence of lipid lowering. *Arterioscler Thromb Vasc Biol* 2002;22(4):692-8.

[81] Ferro D, Basili S, Alessandri C, Mantovani B, Cordova C, Violi F. Simvastatin reduces monocyte-tissue-factor expression type IIa hypercholesterolaemia. *Lancet* 1997;350(9086):1222.

[82] Holschermann H, Hilgendorff A, Kemkes-Matthes B, Schonburg M, Bauer EP, Tillmanns H, et al. Simvastatin attenuates vascular hypercoagulability in cardiac transplant recipients. *Transplantation* 2000;69(9):1830-6.

[83] Cortellaro M, Cofrancesco E, Arbustini E, Rossi F, Negri A, Tremoli E, et al. Atorvastatin and thrombogenicity of the carotid atherosclerotic plaque: the ATROCAP study. *Thromb Haemost* 2002;88(1):41-7.

[84] Hansen JB, Huseby KR, Huseby NE, Sandset PM, Hanssen TA, Nordoy A. Effect of cholesterol lowering on intravascular pools of TFPI and its anticoagulant potential in type II hyperlipoproteinemia. *Arterioscler Thromb Vasc Biol* 1995;15(7):879-85.

[85] Lorena M, Perolini S, Casazza F, Milani M, Cimminiello C. Fluvastatin and tissue factor pathway inhibitor in type IIA and IIB hyperlipidemia and in acute myocardial infarction. *Thromb Res* 1997;87(4):397-403.

[86] Nordoy A, Bonaa KH, Sandset PM, Hansen JB, Nilsen H. Effect of omega-3 fatty acids and simvastatin on hemostatic risk factors and postprandial hyperlipemia in patients with combined hyperlipemia. *Arterioscler Thromb Vasc Biol* 2000;20(1):259-65.

[87] Morishita E, Asakura H, Saito M, Yamazaki M, Ontachi Y, Mizutani T, et al. Elevated plasma levels of free-form of TFPI antigen in hypercholesterolemic patients. *Atherosclerosis* 2001;154(1):203-12.

[88] Krysiak R, Okopien B, Herman Z. Effects of HMG-CoA reductase inhibitors on coagulation and fibrinolysis processes. *Drugs* 2003;63(17):1821-54.

[89] Szczeklik A, Musial J, Undas A, Gajewski P, Gora P, Swadzba J, et al. Inhibition of thrombin generation by simvastatin and lack of additive effects of aspirin in patients with marked hypercholesterolemia. *J Am Coll Cardiol* 1999;33(5):1286-93.

[90] Musial J, Undas A, Undas R, Brozek J, Szczeklik A. Treatment with simvastatin and low-dose aspirin depresses thrombin generation in patients with coronary heart disease and borderline-high cholesterol levels. *Thromb Haemost* 2001;85(2):221-5.

[91] Aoki I, Aoki N, Kawano K, Shimoyama K, Maki A, Homori M, et al. Platelet-dependent thrombin generation in patients with hyperlipidemia. *J Am Coll Cardiol* 1997;30(1):91-6.

[92] Alessandri C, Basili S, Maurelli M, Bracaglia D, Andreozzi P, Pergolini M, et al. Relationship between prothrombin activation fragment F1 + 2 and serum cholesterol. *Haemostasis* 1996;26(4):214-9.

[93] Puccetti L, Bruni F, Di Renzo M, Bova G, Cercignani M, Iadanza A, et al. Hypercoagulable state in hypercholesterolemic subjects assessed by platelet-dependent thrombin generation: in vitro effect of cerivastatin. *Eur Rev Med Pharmacol Sci* 1999;3(5):197-204.

[94] Puccetti L, Bruni F, Bova G, Cercignani M, Palazzuoli A, Console E, et al. Effect of diet and treatment with statins on platelet-dependent thrombin generation in hypercholesterolemic subjects. *Nutr Metab Cardiovasc Dis* 2001;11(6):378-87.

[95] Meade TW, Mellows S, Brozovic M, Miller GJ, Chakrabarti RR, North WR, et al. Haemostatic function and ischaemic heart disease: principal results of the Northwick Park Heart Study. *Lancet* 1986;2(8506):533-7.

[96] Wilhelmsen L, Svardsudd K, Korsan-Bengtsen K, Larsson B, Welin L, Tibblin G. Fibrinogen as a risk factor for stroke and myocardial infarction. *N Engl J Med* 1984;311(8):501-5.

[97] Kannel WB, Wolf PA, Castelli WP, D'Agostino RB. Fibrinogen and risk of cardiovascular disease. The Framingham Study. *Jama* 1987;258(9):1183-6.

[98] Davidson M, McKenney J, Stein E, Schrott H, Bakker-Arkema R, Fayyad R, et al. Comparison of one-year efficacy and safety of atorvastatin versus lovastatin in primary hypercholesterolemia. Atorvastatin Study Group I. *Am J Cardiol* 1997;79(11):1475-81.

[99] Seeger H, Wallwiener D, Mueck AO. Lipid-independent effects of an estrogen-statin combination: inhibition of expression of adhesion molecules and plasminogen activator inhibitor--I in human endothelial cell cultures. *Climacteric* 2001;4(3):209-14.

[100] Warkentin TE. Hemostasis and atherosclerosis. *Can J Cardiol* 1995;11 Suppl C:29C-34C.

[101] Essig M, Nguyen G, Prie D, Escoubet B, Sraer JD, Friedlander G. 3-Hydroxy-3-methylglutaryl coenzyme A reductase inhibitors increase fibrinolytic activity in rat aortic endothelial cells. Role of geranylgeranylation and Rho proteins. *Circ Res* 1998;83(7):683-90.

[102] Mussoni L, Banfi C, Sironi L, Arpaia M, Tremoli E. Fluvastatin inhibits basal and stimulated plasminogen activator inhibitor 1, but induces tissue type plasminogen activator in cultured human endothelial cells. *Thromb Haemost* 2000;84(1):59-64.

[103] Haslinger B, Goedde MF, Toet KH, Kooistra T. Simvastatin increases fibrinolytic activity in human peritoneal mesothelial cells independent of cholesterol lowering. *Kidney Int* 2002;62(5):1611-9.

[104] Lopez S, Peiretti F, Bonardo B, Juhan-Vague I, Nalbone G. Effect of atorvastatin and fluvastatin on the expression of plasminogen activator inhibitor type-1 in cultured human endothelial cells. *Atherosclerosis* 2000;152(2):359-66.

[105] Bourcier T, Libby P. HMG CoA reductase inhibitors reduce plasminogen activator inhibitor-1 expression by human vascular smooth muscle and endothelial cells. *Arterioscler Thromb Vasc Biol* 2000;20(2):556-62.

[106] Ishibashi T, Nagata K, Ohkawara H, Sakamoto T, Yokoyama K, Shindo J, et al. Inhibition of Rho/Rho-kinase signaling downregulates plasminogen activator inhibitor-1 synthesis in cultured human monocytes. *Biochim Biophys Acta* 2002;1590(1-3):123-30.
[107] Wiesbauer F, Kaun C, Zorn G, Maurer G, Huber K, Wojta J. HMG CoA reductase inhibitors affect the fibrinolytic system of human vascular cells in vitro: a comparative study using different statins. *Br J Pharmacol* 2002;135(1):284-92.
[108] Randomised trial of cholesterol lowering in 4444 patients with coronary heart disease: the Scandinavian Simvastatin Survival Study (4S). *Lancet* 1994;344(8934):1383-9.
[109] Sacks FM, Pfeffer MA, Moye LA, Rouleau JL, Rutherford JD, Cole TG, et al. The effect of pravastatin on coronary events after myocardial infarction in patients with average cholesterol levels. Cholesterol and Recurrent Events Trial investigators. *N Engl J Med* 1996;335(14):1001-9.
[110] Prevention of cardiovascular events and death with pravastatin in patients with coronary heart disease and a broad range of initial cholesterol levels. The Long-Term Intervention with Pravastatin in Ischaemic Disease (LIPID) Study Group. *N Engl J Med* 1998;339(19):1349-57.
[111] Stenestrand U, Wallentin L. Early statin treatment following acute myocardial infarction and 1-year survival. *Jama* 2001;285(4):430-6.
[112] Aronow HD, Topol EJ, Roe MT, Houghtaling PL, Wolski KE, Lincoff AM, et al. Effect of lipid-lowering therapy on early mortality after acute coronary syndromes: an observational study. *Lancet* 2001;357(9262):1063-8.
[113] Cannon CP, McCabe CH, Bentley J, Braunwald E. Early statin therapy is associated with markedly lower mortality in patients with acute coronary syndromes: observations from OPUS-TIMI 16 [abstract]. *J Am Coll Cardiol* 2001;37(suppl A):334A. Abstract 831-2.
[114] Newby LK, Kristinsson A, Bhapkar MV, Aylward PE, Dimas AP, Klein WW, et al. Early statin initiation and outcomes in patients with acute coronary syndromes. *Jama* 2002;287(23):3087-95.
[115] Heeschen C, Hamm CW, Laufs U, Snapinn S, Bohm M, White HD. Withdrawal of statins increases event rates in patients with acute coronary syndromes. *Circulation* 2002;105(12):1446-52.
[116] Schiele R, Gitt AK, Heer T. Early statin use in acute myocardial is associated with a reduced hospital mortality: results of the Mitra-2 [abstract]. *Circulation* 2000;102(suppl 2):II-435. Abstract 2117.
[117] Briguori C, Colombo A, Airoldi F, Violante A, Focaccio A, Balestrieri P, et al. Statin administration before percutaneous coronary intervention: impact on periprocedural myocardial infarction. *Eur Heart J* 2004;25(20):1822-8.
[118] Lenderink T, Boersma E, Gitt AK, Zeymer U, Wallentin L, Van de Werf F, et al. Patients using statin treatment within 24 h after admission for ST-elevation acute coronary syndromes had lower mortality than non-users: a report from the first Euro Heart Survey on acute coronary syndromes. *Eur Heart J* 2006;27(15):1799-1804.
[119] Kayikcioglu M, Turkoglu C, Kultursay H, Evrengul H, Can L. The short term results of combined use of pravastatin with thrombolytic therapy in acute myocardial infarction [abstract]. *Circulation* 1999;100(suppl 1):I-303. Abstract 1586.
[120] Schwartz GG, Olsson AG, Ezekowitz MD, Ganz P, Oliver MF, Waters D, et al. Effects of atorvastatin on early recurrent ischemic events in acute coronary syndromes: the MIRACL study: a randomized controlled trial. *Jama* 2001;285(13):1711-8.

[121] Liem AH, van Boven AJ, Veeger NJ, Withagen AJ, Robles de Medina RM, Tijssen JG, et al. Effect of fluvastatin on ischaemia following acute myocardial infarction: a randomized trial. *Eur Heart J* 2002;23(24):1931-7.

[122] Arntz HR, Agrawal R, Wunderlich W, Schnitzer L, Stern R, Fischer F, et al. Beneficial effects of pravastatin (+/-colestyramine/niacin) initiated immediately after a coronary event (the randomized Lipid-Coronary Artery Disease [L-CAD] Study). *Am J Cardiol* 2000;86(12):1293-8.

[123] Serruys PW, de Feyter P, Macaya C, Kokott N, Puel J, Vrolix M, et al. Fluvastatin for prevention of cardiac events following successful first percutaneous coronary intervention: a randomized controlled trial. *Jama* 2002;287(24):3215-22.

[124] Cannon CP, Braunwald E, McCabe CH, Rader DJ, Rouleau JL, Belder R, et al. Intensive versus moderate lipid lowering with statins after acute coronary syndromes. *N Engl J Med* 2004;350(15):1495-504.

[125] Thompson PL, Meredith I, Amerena J, Campbell TJ, Sloman JG, Harris PJ. Effect of pravastatin compared with placebo initiated within 24 hours of onset of acute myocardial infarction or unstable angina: the Pravastatin in Acute Coronary Treatment (PACT) trial. *Am Heart J* 2004;148(1):e2.

[126] de Lemos JA, Blazing MA, Wiviott SD, Lewis EF, Fox KA, White HD, et al. Early intensive vs a delayed conservative simvastatin strategy in patients with acute coronary syndromes: phase Z of the A to Z trial. *Jama* 2004;292(11):1307-16.

[127] LaBlanche JM, Jukema JW, Charbonneau F, Creplet J, Eldar M, Galve E, et al. PRINCESS: Prevention of ischemic events by early treatment of cerivastatin after acute myocardial infarction. In: *ESC Congress*; 2004; *Munich*; 2004.

[128] Ostadal P, Alan D, Vejvoda J. Statins in the first-line therapy of acute coronary syndromes – similar to aspirin? *Exp Clin Cardiol* 2005;10(1):9-16.

[129] Lefer AM, Campbell B, Shin YK, Scalia R, Hayward R, Lefer DJ. Simvastatin preserves the ischemic-reperfused myocardium in normocholesterolemic rat hearts. *Circulation* 1999;100(2):178-84.

[130] Ueda Y, Kitakaze M, Komamura K, Minamino T, Asanuma H, Sato H, et al. Pravastatin restored the infarct size-limiting effect of ischemic preconditioning blunted by hypercholesterolemia in the rabbit model of myocardial infarction. *J Am Coll Cardiol* 1999;34(7):2120-5.

[131] Lefer DJ, Scalia R, Jones SP, Sharp BR, Hoffmeyer MR, Farvid AR, et al. HMG-CoA reductase inhibition protects the diabetic myocardium from ischemia-reperfusion injury. *Faseb J* 2001;15(8):1454-6.

[132] Di Napoli P, Antonio Taccardi A, Grilli A, Spina R, Felaco M, Barsotti A, et al. Simvastatin reduces reperfusion injury by modulating nitric oxide synthase expression: an ex vivo study in isolated working rat hearts. *Cardiovasc Res* 2001;51(2):283-93.

[133] Bauersachs J, Galuppo P, Fraccarollo D, Christ M, Ertl G. Improvement of left ventricular remodeling and function by hydroxymethylglutaryl coenzyme a reductase inhibition with cerivastatin in rats with heart failure after myocardial infarction. *Circulation* 2001;104(9):982-5.

[134] Hayashidani S, Tsutsui H, Shiomi T, Suematsu N, Kinugawa S, Ide T, et al. Fluvastatin, a 3-hydroxy-3-methylglutaryl coenzyme a reductase inhibitor, attenuates left ventricular remodeling and failure after experimental myocardial infarction. *Circulation* 2002;105(7):868-73.

[135] Bell RM, Yellon DM. Atorvastatin, administered at the onset of reperfusion, and independent of lipid lowering, protects the myocardium by up-regulating a pro-survival pathway. *J Am Coll Cardiol* 2003;41(3):508-15.
[136] Wolfrum S, Dendorfer A, Schutt M, Weidtmann B, Heep A, Tempel K, et al. Simvastatin acutely reduces myocardial reperfusion injury in vivo by activating the phosphatidylinositide 3-kinase/Akt pathway. *J Cardiovasc Pharmacol* 2004;44(3):348-55.
[137] Ostadal P, Alan D, Hajek P, Horak D, Vejvoda J, Trefanec J, et al. The effect of early treatment by cerivastatin on the serum level of C-reactive protein, interleukin-6, and interleukin-8 in the patients with unstable angina and non-Q-wave myocardial infarction. *Mol Cell Biochem* 2003;246(1-2):45-50.
[138] Ostadal P, Alan D, Hajek P, Vejvoda J, Mates M, Blasko P, et al. Fluvastatin in the therapy of acute coronary syndrome: Rationale and design of a multicenter, randomized, double-blind, placebo-controlled trial (The FACS Trial) [ISRCTN81331696]. *Curr Control Trials Cardiovasc Med.* 2005;6(1):4.

In: Statins: Indications and Uses, Safety and Modes of Action ISBN 978-1-60692-103-6
Editor: Gunnar N. Holmquist

Chapter 7

BAD HDL-C RESPONDERS TO STATINS

Dirk Devroey*
Department of General Practice, Dutch-Speaking University of Brussels, Belgium

ABSTRACT

Lowering high levels of low-density lipoprotein cholesterol (LDL-C) is the primary aim in the prevention of cardiac events. However, low levels of high-density lipoprotein cholesterol (HDL-C) are also associated with an increased risk of ischemic heart disease.

Lipid-lowering drugs are known to decrease LDL-C and to increase HDL-C slightly. However, not all patients benefit from this effect. Some patients have lower HDL-C during statin treatment than before the treatment.

These patients were first described in a case report in 2002 as 'bad HDL-C responders to statins'. In the case of one man, HDL-C and the ratios of total cholesterol (TC) to HDL-C and LDL-C to HDL-C worsened dramatically during pravastatin treatment. After 3 years, pravastatin was replaced by fenofibrate. The result was spectacular. The HDL-C increased to at least twice the level obtained during pravastatin.

Bad HDL-C responders are characterized by HDL-C levels which decrease below 40 mg/dl during the treatment, despite higher HDL-C levels before the treatment.

The existence of bad HDL-C responders to statins was confirmed by a prospective survey of 2,259 patients treated with a statin or a fibrate for hyperlipidaemia. The proportion of bad HDL-C responders is higher for statins (6%) than for fibrates (4%).

In a review of the guidelines, almost all selected guidelines consider low HDL-C as a marker of increased risk for coronary heart disease. However, only few guidelines use the level of HDL-C as a threshold or target level for the treatment of dyslipidemia. The guidelines provide only little information on the management of patients with treatment-induced low HDL-C. Instead of using TC or LDL-C we consider the use of the ratios of TC to HDL-C or LDL-C to HDL-C as a threshold as well as a target for treatment.

---

* Address correspondence to: Dr. Dirk Devroey, University of Brussels (VUB), Dept of General Practice, Laarbeeklaan 103, 1090 Brussels, BELGIUM, dirk.devroey@vub.ac.be

Treatment with fibrates was studied in 14 bad HDL-C responders to statins. Far better levels for HDL-C, TC to HDL-C and LDL-C to HDL-C were obtained with fibrates compared to statins. For bad HDL-C responders to statins with low or normal LDL-C, treatment with fibrates instead of statins should be considered. For those with high LDL-C, fibrates should be added to statins.

Treatment for bad HDL-C responders should be studied in randomized controlled trials. Such a trial with simvastatin and fenofibrate has been initiated to corroborate our findings.

Keywords: dyslipidemia, high-density lipoprotein cholesterol, antilipemic agents, hypercholesterolemia, lipids.

Introduction

The reduction of low-density lipoproteins (LDL-C) is the primary approach to decrease the risk of coronary heart disease (CHD) in both primary and secondary prevention [1]. In the mid-nineties, the results of the statin trials created the belief that LDL-C was the most important lipid fraction in the prevention of CHD [2-6]. However, the Adult Treatment Panel III (ATP-III) sets HDL-C levels of < 40 mg/dl as a categorical risk factor and designates it as a factor that modifies the LDL-C goal [1]. Some studies have found that low levels of high-density lipoprotein cholesterol (HDL-C) are also strongly associated with an increased risk of CHD [7-8].

According to the ATP-III guidelines non-HDL cholesterol is the secondary target in the treatment of hypercholesterolemia. Once adequate LDL-C levels have been obtained, other lipid risk factors such as low HDL-C deserve attention.

Since the nineteen fifties, epidemiologic studies had found that low levels of HDL-C are strongly associated with an increased risk of CHD. Already in 1951, the relation between low levels of HDL-C and CHD was noticed [9]. In 1986, the Framingham study confirmed the association between low HDL-C and CHD [10].

For subjects with a history of CHD, low HDL-C is at least as prevalent as high LDL-C [11]. The role of HDL-C and triglycerides (TG) was highlighted as therapeutic targets for the prevention and treatment of CHD [12].

Lipid-lowering drugs are known to decrease total cholesterol (TC) and LDL-C and to increase HDL-C slightly. They consequently lower morbidity and mortality by CHD in primary and secondary prevention [3,4,13]. However, some patients show a similar decrease for HDL-C as for TC and LDL-C resulting in lower HDL-C levels during the treatment than before the initiation of the treatment. This sometimes results in higher ratios of TC/HDL-C and LDL-C/HDL-C during the treatment with the lipid-lowering drugs than before it.

This phenomenon was first described in a case report and was called "bad HDL-C response to statins" [14]. Bad HDL-C responders were defined as patients whose HDL-C level decreases below 40 mg/dl during treatment with lipid-lowering drugs, despite higher HDL-C levels before the treatment. The case report describes a man whose HDL-C and the ratios of TC to HDL-C and LDL-C to HDL-C worsened dramatically during pravastatin treatment. After 3 years, pravastatin was replaced by fenofibrate. The result was spectacular. The HDL-C increased to at least twice the level obtained during pravastatin [14].

Although the effect of fenofibrate is quite convincing in this case report, physicians need more evidence than a case report before treatment guidelines for these patients can be adapted. During the course of our research we tried (1) to confirm the existence of bad HDL-C responders, (2) to determine the causes for bad HDL-C response, and (3) to determine the prevalence of bad HDL-C response. Finally the guidelines were reviewed on the treatment of low HDL-C and suggestions were made for a more adequate treatment of these patients.

The *existence* of bad HDL-C responders to statins was confirmed by a prospective survey of 2,259 representative patients treated with a statin or a fibrate for hyperlipidaemia [15].

The *prevalence* of low HDL-C (< 40 mg/dl) was determined before and during lipid-lowering treatment [16]. Additionally the prevalence of low HDL-C during fibrate and statin treatment was compared.

In a review of the treatment *guidelines*, we aimed to determine the importance given to HDL-C as risk factor or as threshold and target level in the treatment of dyslipidemia [17]. We developed a strategy with cholesterol-related keywords to search for guidelines on the major databases. The AGREE instrument was used for the evaluation and inclusion of the guidelines.

Treatment with fibrates was studied among 14 bad HDL-C responders to statins [18]. The aim was to describe the benefit of fibrates in monotherapy for these patients. It was studied in a cross-sectional survey including lipid levels, cardiovascular disease and risk factors.

The First Case Report of a Bad HDL-C Responder to Statins

The first case report of a bad HDL-C responder to statins concerned a male born in 1942. He had been treated for essential hypertension since 1989. He had no family history of cardiovascular disease. At that time he consumed a moderate-fat diet and exercised a few times each week. His body mass index (BMI) was 33.9 kg/m^2. On a fasting lipid panel during 1992, values were as follows: TC at 232 mg/dl, LDL-C at 168 mg/dl, HDL-C at 42 mg/dl, and TG at 101 mg/dl. He was advised a low-fat diet. No hypolipidemic drugs were prescribed.

In September 1995, he developed a stroke with right motor deficit and speech problems. The treatment at the discharge from hospital consisted of atenolol (50 mg OD), felodipine (5 mg OD) and acetylsalicylic acid (100 mg OD). After 1 year the motor functions recovered almost completely. He could resume his daily activities, including a walk of about 1 h.

In October 1996, fasting blood levels showed TC at 248 mg/dl, HDL-C at 43 mg/dl, LDL-C at 162 mg/dl and TG raised to 215 mg/dl. Micronised fenofibrate (200 mg OD) was added to the treatment.

Three months later, a fasting blood panel showed considerably better lipid levels with TC at 181 mg/dl, HDL-C at 55 mg/dl, LDL-C at 115 mg/dl and TG levels at 54 mg/dl.

Because of the promising results of the statin trials in secondary prevention, fenofibrate was switched to pravastatin (20 mg OD) in March 1997. Three months later, a fasting blood panel showed TC levels at 185 mg/dl with HDL-C at 42 mg/dl, LDL-C at 118 mg/dl and TG at 124 mg/dl. At that time, atenolol was switched to losartan (50 mg OD) because of bradycardia.

Lipid levels remained stable for about 30 months. Only the HDL-C levels decreased during this period to finally 24 mg/dl in July 1999. The TC/HDL-C ratio increased from 3.3 in 1997 to 7.4 in 1999 (Figure 2). During the period concerned, no drugs influencing lipid levels such as beta-blockers, diuretics or hypoglycemic drugs were added to the treatment.

Because the TC/HDL-C ratio was far better during fenofibrate treatment, this drug was started up again to replace pravastatin. The result was spectacular. The HDL-C increased within 3 months to 62 mg/dl and the TC/HDL-C ratio decreased to 2.9. This favorable result persisted during the following years up till today. The result was not related to changes in life-style, diet, drugs or physical exercise.

We certainly do not want to put in doubt the importance of statins in secondary prevention but we consider the existence of "bad HDL-C responders" who can be defined as subjects whose HDL-C levels decrease dramatically during statin treatment.

For the patient of this case report, the difference in response between pravastatin and fenofibrate cannot be explained by changes in co-medication, diet, life-style or physical exercise. The reason for a bad response to statins remains unclear, but it has to be borne in mind that (1) the different response to fibrates and statins with regard to HDL-C is not surprising in view of their different biological action mechanisms, and (2) the genetic makeup may significantly influence the way patients respond to a cholesterol-lowering drug.

One could argue that for the bad responders a fibrate should be added to the statin. But for this patient a sufficient result was obtained with a fibrate in monotherapy.

In conclusion, far better levels of HDL-C, TC/HDL-C and LDL-C/HDL-C were achieved with fenofibrate in this patient with low HDL-C during pravastatin treatment, and a favorable effect on CHD-related morbidity and mortality can be expected. Before generalizing, these findings should be confirmed by a cohort study or a randomized clinical trial.

Confirmation about the Existence of Bad HDL-C Responders

Once the first case of a bad HDL-C responder to statins was described, we tried to identify similar patients in our practice. Among our patients who experienced a major cardiovascular event before or during statin treatment, five had HDL-C levels below 30 mg/dl. After switching these patients from a statin to a fibrate, the HDL-C level almost doubled. No evidence about this was found in the literature. However, micronized fenofibrate has shown to significantly increase mean HDL-C levels to above 40 mg/dl, irrespective of baseline HDL-C levels [19]. The VA-HIT was the first double-blind trial demonstrating that raising the HDL-C with a fibrate resulted in a significant reduction in the risk of major cardiovascular events in patients with CHD whose primary lipid abnormality was a low HDL-C level associated with an average LDL-C level [20].

In almost all clinical trials the lipid levels are represented by medians and standard deviations and by the changes from baseline in terms of percentages. The results are rarely represented by percentages of patients achieving the target levels for TC, LDL-C, HDL-C and TG. For that reason we analysed all our patients treated for hyperlipidaemia retrospectively. The aim of this analysis was to compare the proportions of patients achieving the HDL-C target levels after one year of treatment with statins or fibrates. Furthermore, a subgroup with low HDL-C levels during statin treatment was investigated and suggestions are made for a better management of these patients.

Data Recording

None of the 120 treated patients had familial hypercholesterolemia. Treatment was started after two independent measurements of TC above 250 mg/dl or TG above 200 mg/dl after one night fasting. The first measurement was done before the start of a diet, the second one after a diet of at least three months. After the first blood test the patients were given oral and written guidelines concerning a lipid lowering diet corresponding to the Step I diet of the American Heart Association [21]. All patients continued their diet during the treatment with fibrates or statins. TC, LDL-C, HDL-C and TG were extracted from the medical records before the start of a diet, after a diet of at least three months and, if applicable, after twelve months of treatment with a statin or a fibrate. None of the patients was treated at the same time with statins and fibrates. No particular motives for selection of a statin or a fibrate at baseline existed. The personal and family medical history for major cardiac events, ischemic heart diseases, hypertension, stroke and diabetes were recorded cross-sectionally.

The proportion of patients achieving the treatment target levels was determined at baseline and after 12 months of treatment with a statin and/or a fibrate. The treatment target levels were 190 mg/dl for TC, 115 mg/dl for LDL-C, 40 mg/dl for HDL-C, 150 mg/dl for TG, 4.0 for TC/HDL-C and 3.0 for LDL-C/HDL-C [1].

For plasma lipid measurements venous blood samples were collected after the subjects had fasted for 12 hours. Samples were transported in Becton Dickinson SST Vacutainer 6 ml glass tubes. All analyses were performed within eight hours after blood sampling at the LBH II laboratories in Lasne-Ohain Belgium. Serum TC, TG and HDL-C were measured enzymatically on a Hitachi 917 analyzer. LDL-C levels were always calculated with the Friedewald formula, unless TG were above 300 mg/dl [22]. In that case LDL-C levels were measured on the Hitachi 917 analyzer.

Risk assessment for determining 10-year risk for CHD was carried out according to the updated Framingham risk scoring [23]. The CHD risk was high for patients with a calculated 10-year risk over 20% and for patients who experienced a cardiac or cerebrovascular event or having diabetes. The risk was medium for a calculated 10-year risk between 10 and 20%. Low risk patients had a calculated 10-year risk below 10%.

SPSS-PC 10® (SPSS Inc.,Chicago,IL,USA) was used for analysis and statistical processing. Significant differences were detected with the independent-samples t-test. The cross-tables from Epi Info 2000® were used to detect differences between groups by means of chi-square tests.

Baseline Values

In total 120 patients receiving a lipid-lowering drug were included. Fifty-three percent were men. The median age was 64 years ranging from 31 to 90 years. Baseline blood tests were only available for 80 of these 120 patients. The other patients had already been treated for dyslipidaemia before the first visit to our practice. The baseline values after a diet of at least three months are shown in tables 1 and 2.

Table 1. Mean lipoproteins (in mg/dl) at baseline and after 12 months of treatment with statins (n=34).

	Baseline	12 months
TC	279	215
LDL-C	195	138
HDL-C	45	51
TG	174	129
TC / HDL-C	6.8	4.7
LDL-C / HDL-C	4.7	3.1

Table 2. Mean lipoproteins (in mg/dl) at baseline and after 12 months of treatment with fibrates (n=46).

	Baseline	12 months
TC	277	213
LDL-C	195	141
HDL-C	46	49
TG	181	114
TC / HDL-C	6.5	4.6
LDL-C / HDL-C	4.5	3.1

Treatment Results

Mean duration of the treatment was 1.2 year for the fibrates and 1.1 year for the statins. In the fibrate group (n=46), the medical treatment consisted of fenofibrate micronized 200 mg OD (80%) or ciprofibrate 100 mg OD (20%). In the statin group (n=34), patients were treated with pravastatin 20 mg OD (56%), simvastatin 20 mg OD (35%), atorvastatin 10 mg OD (3%) or fluvastatin 40 mg OD (6%).

The mean levels of TC, LDL-C, TG, TC/HDL-C and LDL-C/HDL-C decreased after twelve months of treatment by 23, 29, 26, 31 and 34%, respectively, for the patients treated with statins and by 22, 28, 37, 29 and 31%, respectively, for the patients treated with fibrates. HDL-C levels increased by 13 and 6% for statins and fibrates, respectively (Tables 1 and 2). No significant differences were observed between both groups.

The number of patients not achieving the treatment target levels was calculated for the statin and the fibrate group. Comparing these proportions, no major differences were detected between both groups concerning TC (25% for statins and 21% for fibrates, respectively), LDL-C (31% and 17%), TG (81% and 65%), TC/HDL-C (35% and 31%) and LDL-C/HDL-C (52% and 42%). But for HDL-C the number of patients achieving the target of 40 mg/dl was higher for the fibrate group (82%) compared to the statin group (62%) ($p<0.042$).

Bad HDL-C Responders

Of the 120 analyzed patients, 65 patients had been treated with statins (34 with known baseline levels and 31 without known baseline levels). After 1.1 year of treatment, 25 (39%) had low HDL-C levels (< 40 mg/dl). The baseline characteristics for these 'bad HDL-C responders' are compared with those of the 'good responders'. The treatment with statins for good and bad HDL-C responders, respectively, consisted of pravastatin 20 mg OD (58% and 48%), simvastatin 20 mg OD (37% and 44%), atorvastatin 10 mg OD (0% and 4%) or fluvastatin 40 mg OD (5% and 4%). There was no difference between simvastatin and pravastatin in terms of good or bad HDL-C responders ($p=0.523$).

In the group of the bad HDL-C responders more men (80% compared to 48% in the good responders; $P=0.01$), lower baseline levels of HDL-C (40 mg/dl compared to 48 mg/dl; $P=0.041$), more myocardial infarctions (32% compared to 3%; $P= 0.001$) and more patients at high coronary risk (over 20%) (76% compared to 50%) were observed. Between good and bad HDL-C responders no significant differences were found concerning age (63 and 61 years, respectively), body mass index (26 and 27 kg/m^2), proportions of obese patients (5% and 8%), overweight patients (55% and 72%), smokers (28% and 36%), diabetes (10% and 12%) stroke (3% and 12%) and hypertension (58% and 68%).

The Confirmation about Bad HDL-C Responders to Statins

According to the baseline characteristics we can conclude that the population involved corresponds with the populations described in other mixed primary and secondary prevention studies. The observed lipid changes from baseline don't differ from the results of most major long-term clinical trials. The statin group is not biased by patients who do not respond well to statins because, compared to baseline, the mean HDL-C changes are similar in the statin and the fibrate group. The number of patients not achieving the HDL-C target level is higher for statins than for fibrates. Because the mean HDL-C similarly increased for both treatment groups, we consider that two groups of responders to statins could exist: one group with an important increased HDL-C and one with a less pronounced or even decreased HDL-C during treatment with statins. This suggests the existence of 'good' and 'bad' HDL-C responders to statins. Probably the 'good' and 'bad HDL-C responders' will not exist as two separate groups since the transition between both groups is more than likely gradual. This can, nevertheless, explain why the mean HDL-C level increases with statins but fewer patients achieve the HDL-C target levels with statins.

Bad HDL-C response to statins doesn't seem to be related to the type of statin used. No differences between the different statins were demonstrated. In our study few patients received atorvastatin because only the first year ever of lipid-lowering treatment was taken into account. At the moment of the extraction of the data from the medical records (December 2000) atorvastatin had been for sale since two years and only few patients had had atorvastatin as the initial treatment for hyperlipidaemia. At that time the recommended dose for simvastatin as well as pravastatin was 20 mg. Shortly after the data extraction the recommended dose was increased to 40 mg.

In this study the majority of the bad HDL-C responders are men, have low baseline HDL-C and have a higher risk for cardiovascular events. Of the 'good' responders, only 3% had a myocardial infarction instead of 32% of the 'bad' responders.

Although it was not the primary objective of our study to evaluate cardiac events, it is noteworthy that the bad HDL-C responders to statins did have a higher incidence of serious cardiovascular events.

The Prevalence of Bad HDL-C Responders

Statins as well as fibrates lower total cholesterol (TC) and LDL-C and increase HDL-C. With statins HDL-C generally rises by 5-10% but greater increases usually occur in patients with low HDL-C and elevated triglycerides (TG) [3,4,24]. Fibrates usually raise HDL-C by 10-15% but greater increases can occur in patients with very low HDL-C levels [25].

Figures about the occurrence of low HDL-C levels before and during lipid-lowering treatment are only available from the treat-to-goal trials but only few epidemiologic figures in real-life conditions are available.

The aim of this study was to estimate the prevalence of low HDL-C (< 40 mg/dl) before and during lipid-lowering treatment. Additionally the prevalence of low HDL-C and the bad HDL-C response to lipid-lowering treatment was compared for fibrates and statins.

Data Collection

To determine the prevalence of bad HDL-C responders the data collection was based on a review of the requests for prolongation of the reimbursement of lipid-lowering drugs in Belgium. During February and March 2002 all requests for prolongation of the reimbursement were recorded at the regional offices of two Health Insurance Associations (Liberale Mutualiteit Brabant and Vlaams Neutraal Ziekenfonds Aalst). The requests were completed by family physicians as well as by specialists.

The following data were available: date of birth, sex, the actual lipid-lowering drug and doses. The most recent fasting lipoprotein levels (TC, TG, HDL-C and LDL-C) and those after a three-month diet before the start of the treatment were recorded. All studied patients received a lipid-lowering drug during at least three months but most of the patients had been treated since several years.

For plasma lipid measurements venous blood samples were collected after the subjects had fasted for 12 hours. Samples were transported in glass serum tubes. Serum TC, TG and HDL-C were measured enzymatically. LDL-C levels were always calculated with the

Friedewald formula, unless TG were above 300 mg/dl [22]. In that case LDL-C levels were measured enzymatically. All tests were performed by local laboratories. According to the Belgian guidelines for clinical biology, performances of these laboratories are regularly subjected to internal as well as external quality control.

Low HDL-C was defined as HDL-C levels below 40 mg/dl. A difference was made between the patients with initially normal levels of HDL-C (> or = 40 mg/dl) presenting low levels of HDL-C (< 40 mg/dl) during the treatment and the patients with initially low levels of HDL-C which remain low despite the treatment. We suggest defining the first as 'bad HDL-C responders' and the second as 'non HDL-C responders'.

Unfortunately no uniform information about cardiovascular risk factors is available from the medical files at the Health Insurance Associations. Only for the few patients with familial hypercholesterolemia were some of the risk factors recorded. For this reason, our study population contains patients in both primary and secondary prevention.

Baseline Values

In total 2259 patients were included. Fifty-six percent were women. The average age was 66 years. The baseline fasting lipoprotein levels after a diet of at least three months were: TC = 286 mg/dl, HDL-C = 61 mg/dl, LDL-C = 184 mg/dl and TG = 209 mg/dl. The baseline fasting lipoprotein levels are shown in table 3 separately for the patients receiving later on statins or fibrates.

Table 3. Average lipoprotein levels (in mg/dl) at baseline and during treatment with statins (n=1548).

	Baseline	Treatment
TC	289	208
LDL-C	196	121
HDL-C	63	58
TG	173	151
TC / HDL-C	5.1	3.9
LDL-C / HDL-C	3.5	2.3

The patients who received fibrates had initially higher TG and lower HDL-C. Those who were prescribed statins had initially higher levels of TC and LDL-C. Low HDL-C levels (< 40 mg/dl) before the initiation of the lipid-lowering drug were found among 9% of the patients (7% for the patients who received later statins and 11% for those who received later fibrates). In total 21% of the men and 4% of the women had low HDL-C before the start of the lipid-lowering drug.

Pharmaceutical Treatment

Sixty-nine percent received statins and 31% fibrates. The average age of the patient treated with fibrates was 68 years (53% women) and for statins 65 years (63% women). The follow-up for both fibrate and statin patients was done in 96% by family physicians and in 4% by specialists. The following drugs were prescribed: atorvastatin (34%), fenofibrate (23%), simvastatin (19%), pravastatin (13%), ciprofibrate (9%) and others drugs such as bezafibrate and fluvastatin in 1%. In total 55% of the patients treated with statins received low doses (atorvastatin 10 mg, pravastatin 20 mg or simvastatin 20 mg) and 45% received high doses (atorvastatin 20 mg, pravastatin 40 mg or simvastatin 40 mg).

Treatment Results

The mean lipoprotein levels during the treatment were: TC = 208 mg/dl, HDL-C = 58 mg/dl, LDL-C = 122 mg/dl and TG = 141 mg/dl. The lipoprotein levels during the treatment are displayed separately for fibrates and statins in table 3. During the treatment 11% of the patients had low HDL-C levels (10% of the statin patients and 13% of the fibrate patients). Eighteen percent of men and 5% of women had low HDL-C (< 40 mg/dl) during the treatment.

Six percent of the patients treated with statins had low HDL-C during the treatment although they had initially normal HDL-C. These patients are the real "bad HDL-C responders to statins".

For fibrates we could also identify patients with initially normal HDL-C levels and low HDL-C during the treatment. Four percent of all patients treated with fibrates had such a bad HDL-C response.

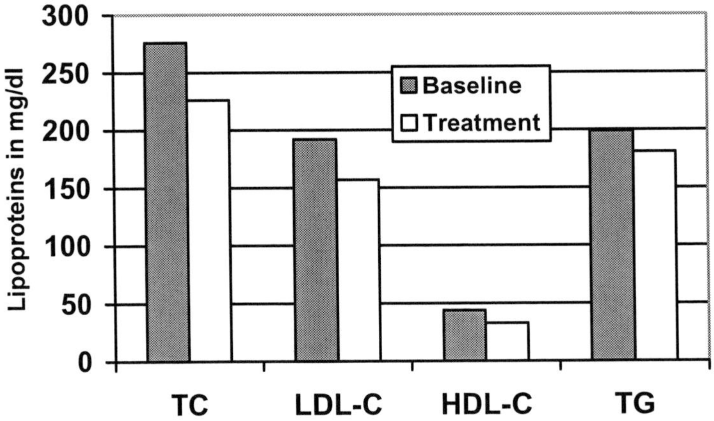

Figure 1. Average lipoprotein levels before and during treatment with fibrates and statins for the bad HDL-C responders only (n=123).

The proportion of bad HDL-C responders to the treatment, in terms of HDL-C levels, which decrease below 40 mg/dl, was higher among the patients treated with statins (6%) than among those treated with fibrates (4%) ($p=0.019$). In total 76% of the bad HDL-C responders to statins and 62% of the bad HDL-C responders to fibrates were men.

Three percent of the patients treated with statins and 2% of those treated with fibrates had initially low levels of HDL-C but turned to have HDL-C levels above 40 mg/dl with the treatment. However, this difference was not significant ($p=0.06$).

No significant differences were observed between the statins mutually in terms of bad HDL-C response. Nor were such differences observed for fibrates.

The ratios of TC to HDL-C and LDL-C to HDL-C increased for the bad HDL-C responders from respectively 6.5 and 4.5 before treatment to 7.0 ($p<0.05$) and 4.9 ($p<0.05$) during treatment (Table 4).

Table 4. Average lipoprotein levels (in mg/dl) at baseline and during treatment with fibrates (n=711).

	Baseline	Treatment
TC	276	208
LDL-C	174	126
HDL-C	59	58
TG	234	121
TC / HDL-C	5.0	3.9
LDL-C / HDL-C	3.3	2.4

Low HDL-C Levels

The proportion of patients with low HDL-C at baseline in our study is comparable with that of the Belgian LIPI-GOAL trial [26]. It concerned a mixed primary (45%) and secondary (55%) treat-to-target study with atorvastatin and 9% of the patients had baseline HDL-C below 40 mg/dl. A Norwegian study demonstrated that in 1997/98 fewer men than women attained the HDL-C target (64% and 86%, respectively) [27]. The prevalence of low HDL-C was lower in our registration but we also found more women than men attaining the HDL-C target.

In the EUROASPIRE low HDL-C levels were found among 20% the treated patients (24% in men and 8% in women) [28]. In this European registration 72% of the patients received statins, 25% fibrates and 3% other lipid-lowering drugs.

In conclusion, the proportion of bad HDL-C responders was 6% for statins and 4% for fibrates. Although lipid-lowering drugs are known to increase the HDL-C levels slightly, not all patients benefit from this effect.

Causes for Bad HDL-C Response

Epidemiological Evidence on Low HDL-C

Low HDL-C levels are prevalent in CHD patients but many of them do not have concomitantly elevated LDL-C. For that reason they are not necessarily considered candidates for lipid-lowering therapy, nor were they represented in the large statin RCTs. Lipid screening of 8500 middle-aged men with CHD found 64% to have HDL-C less than 40 mg/dl and 41% of patients with LDL-C below 130 mg/dl had HDL-C less than 35 mg/dl [11].

Most epidemiological studies highlight the importance of low HDL-C as an independent risk factor for CHD. Convincing epidemiologic evidence of the association of low HDL-C with CHD was derived from the Framingham Heart Study [29]. In this study 5000 subjects initially free from apparent CHD were followed longitudinally for 30 years. The majority of subsequent myocardial infarctions occurred in patients with mildly increased TC (200 to 250 mg/dl). Low HDL-C was identified as the strongest lipid predictor of CHD risk, with levels less than 35 mg/dl associated with an eightfold greater risk than levels over 65 mg/dl.

The Munster Heart Study (PROCAM) was initiated to examine cardiovascular risk factors and cardiovascular events in 17437 men and 8065 women at work [30]. Elevated TC, LDL-C and TG and low HDL-C showed a significant ($P < 0.001$) age-adjusted correlation with the presence of major coronary events.

A combined analysis of four prospective studies including more than 15,000 patients demonstrated a strong and independent link between HDL-C levels and CHD risk, such that a 1 mg/dl increment in HDL-C was associated with an average 2 to 3% decrement in the rate of CHD events [31].

In general, statins increase HDL-C less than fibrates but the absolute HDL-C increase is relatively small with both drug classes [3,4,7,21]. Some occasional paradoxical HDL-C responses in individual patients were observed earlier for fibrates but less for statins.

Bad HDL-C response to statins doesn't seem to be related to the type of statin or fibrate used. No differences between the different statins or fibrates were demonstrated.

Causes for Low HDL-C

Secondary to Modifiable Risk Factors

Low HDL-C can be secondary to modifiable risk factors. Cigarette smoking, obesity, lack of physical activity, higher carbohydrate levels of the diet and smoking tend to be associated with low HDL-C.

In underdeveloped countries, such as Ghana and the Philippines, HDL-C and TC are lower than in countries such as The Netherlands, Finland and Italy [32]. This is related to the lower intake of total and saturated fat in underdeveloped countries.

Some Diseases

Some diseases such as hypothyroidism, obstructive liver diseases, nephritic syndrome and chronic renal failure can decrease HDL-C levels.

Some Drugs

Some drugs such as beta-blockers, androgens and progestagens can also potentially influence HDL-C levels.

There are reasons to believe that in some cases low HDL-C could be related to the intake of statins. Statins are known to decrease TC and LDL-C and to slightly increase HDL-C. Some patients, however, show a decrease of HDL-C as well as LDL-C resulting in lower HDL-C levels during the treatment with a statin than before the initiation of the treatment. This can result in higher ratios of TC to HDL-C and LDL-C to HDL-C during the treatment with a statin than before.

Possible Causes for bad HDL-C Response

Compliance to the Diet

The decrease of HDL-C during the treatment may be related partly to the differences in compliance for the diet between fibrate and statin patients. Statins have more important influences on the lipoprotein levels than fibrates [3,4,24,25]. This may be the reason why statin patients are less compliant to their diet than those receiving fibrates. The lack of compliance to a diet can result in lower levels of HDL-C.

Myalgia

A less probable hypothesis is the one that myalgia among patients receiving lipid-lowering drugs could result in physical inactivity and a consequently decrease of HDL-C. Moreover, myalgia seems to occur more often among patients treated with statins than among those treated with fibrates [1].

Metabolic Pathway

A more probable hypothesis for the higher proportion of bad HDL-C responders with statins compared to fibrates could be related to differences in the metabolic pathway and action between fibrates and statins. The switch of the bad HDL-C responders from statins to fibrates can result in increased levels of HDL-C with more favorable TC to HDL-C and LDL-C to HDL-C ratios [14].

Genetic Profile

The genetic profile of the patient can also play an important role. The known polymorphisms in the beta-fibrinogen gene, the lipoprotein lipase gene and the hepatic lipase gene have an influence on the effects of statins in the general population [33]. The expectation is that in the future a subject's genotype may determine whether he will be treated with statins or fibrates.

Interindividual variability in LDL-C response during treatment with statins is well documented. Poor LDL-C responders to statins have a low basal rate of cholesterol synthesis that may be secondary to a genetically determined increase in cholesterol absorption, possibly mediated by apolipoprotein E4 [34].

According to the Turkish Heart Study, Turks have the lowest levels of HDL-C ever reported [35]. Turks living in Turkey, Germany, and the United States have similarly low HDL-C and all have elevated hepatic lipase activity, which are arguments for a genetic cause for low HDL-C [36].

The Guidelines about Low HDL-C

Current guidelines recommend statins in the prevention of coronary heart disease (CHD) for subjects with high lipid levels [1]. The statin trials created the belief that LDL-C was the most important lipid fraction in the prevention of CHD [5]. However, already in 1986, the Framingham study revealed the association between low HDL-C and CHD [10]. Recently, the Veterans Affairs-High Density Lipoprotein Intervention Trial (VA-HIT) demonstrated the impact of increasing low HDL-C on cardiovascular events [20].

The role of low HDL-C as a risk factor is now more or less accepted. The importance of HDL-C is reflected in some cardiovascular risk scores. However, low HDL-C as a threshold or a target level for treatment seems to be poorly implemented.

Little information is available about the management of low HDL-C. The aim of this study is to describe the importance given to HDL-C as risk factor as well as threshold or target level in the most important diagnostic and treatment guidelines. Additionally, the recommendations for patients with treatment-induced low HDL-C were analyzed in the selected guidelines.

Methods of the Review

The literature search for this review was conducted with use of the major cholesterol-related keywords: lipoproteins, cholesterol, HDL cholesterol, hyperlipidemia, consensus development, practice guidelines, guidelines, drug therapy, antilipemic agents, coronary atherosclerosis, coronary disease, cerebrovascular disorders, outcomes and process assessment. The search was limited to adults on the one hand and to guidelines published between 1997 and 2002 on the other hand. If updates were available, only the most recent version was selected. Translations of existing guidelines were excluded.

With the above keywords we developed a strategy to search for guidelines on Medline, Embase, Cochrane Library, Clinical Evidence and National Guideline Clearinghouse. The search was supplemented with the guidelines found in the bibliographies of the selected guidelines.

The Appraisal of Guidelines Research Evaluation (AGREE) instrument was used for the evaluation and inclusion of the guidelines. The AGREE instrument is based on an appraisal tool developed and validated in the United Kingdom [37]. It was tested on 86 clinical guidelines from ten European countries and Canada [38]. It has already been used by several researchers, for example to compare guidelines on diabetes [39].

Two assessors (Karolien Vantomme and Dirk Devroey) independently evaluated the data. The guidelines were evaluated on their scope and purpose, on stakeholder involvement, on rigor of development, on clarity and presentation, on applicability and on editorial independence. Disagreements between the two assessors were resolved by consensus. Guidelines that scored below 40% in at least one of the six appraisal domains were not included in the review.

In the data assessment a clear difference was made between threshold and target levels. Threshold levels are the values at which a treatment is initiated. Target levels are the levels which are targeted during treatment.

Selected Guidelines

Nine guidelines were selected: National Cholesterol Education Program, Adult Treatment Panel III (ATPIII) [1], Second Joint Task Force of European Societies on Coronary Prevention (SJTF) [40], Dutch College of General Practitioners' practice guideline on lipid-lowering treatment (NHG) [41], Finish Medical Society Duodecim (FMS) [42], Scottish Intercollegiate Guidelines Network (SIGN) [43], American Association of Clinical Endocrinologists (AACE) [44], University of Michigan Health System (UMHS) [45], Institute for Clinical System Improvement (ICSI) [46], and International Task Force for Prevention of Coronary Heart Disease (ITF) [47]. Table 5 shows the role of HDL-C as threshold and target levels in primary and secondary prevention for all selected guidelines.

Table 5. Role of high-density lipoprotein cholesterol as threshold or target level in the management of dyslipidemia.

	Threshold level		Target level	
	Secondary prevention	**Primary prevention**	**Secondary prevention**	**Primary prevention**
SJTF	Risk Factor	Risk Factor	No Risk Factor	No Risk Factor
NHG	No Risk Factor	TC/HDL-C	No Risk Factor	No Risk Factor
FMS	No Risk Factor	No Risk Factor	No Risk Factor	No Risk Factor
SIGN	Risk Factor	-	No Risk Factor	-
ATPIII	Risk Factor	Risk Factor	No Risk Factor	No Risk Factor
AACE	Risk Factor HDL-C	Risk Factor HDL-C	Risk Factor HDL-C	Risk Factor HDL-C
UMHS	Risk Factor HDL-C	Risk Factor HDL-C	Risk Factor HDL-C	Risk Factor HDL-C
ICSI	Risk Factor	Risk Factor	HDL-C	No Risk Factor
ITF	Risk Factor	Risk Factor	Only Diabetics	Only Diabetics

Risk Factor = HDL-C is considered as a risk factor;
No Risk Factor = HDL-C levels are not taken into account as risk factor;
HDL-C = HDL-C level is taken into account;
TC/HDL-C = the ratio of total cholesterol to high-density lipoprotein cholesterol is taken into account;
Only Diabetics = HDL-C is considered for diabetes patients only

Diagnostic Threshold Levels

HDL-C is taken into account as a diagnostic threshold for dyslipidemia in the UMHS and AACE guidelines. HDL-C is considered as a marker of increased cardiovascular risk in the SJTF, SIGN, ATPIII, AACE, UMHS, ICSI and ITF guidelines. However, these guidelines don't include HDL-C in the coronary risk tables.

In the NHG guidelines the ratio of TC to HDL-C is considered as a threshold for treatment in primary prevention. The SJTF guidelines agree that the use of the ratio of TC to HDL-C improves the CHD risk prediction. However, the ratio is not used because HDL-C is not routinely measured in every European country. In the FMS guidelines HDL-C levels are not taken into account at all.

Although low HDL-C is generally considered as a marker for increased risk, most of the risk calculation tables don't take HDL-C levels into account. The ATPIII guidelines are the only ones to incorporate HDL-C in the calculation of coronary risk. ATPIII sets HDL-C levels of < 40 mg/dl as a categorical risk factor and designates it as a factor that modifies the LDL-C goal. Some guidelines use different HDL-C threshold levels for both genders. The ITF guidelines use 35 mg/dl for men and 40 mg/dl for women. The AACE guidelines use respectively 35 and 45 mg/dl.

Treatment Target Levels

HDL-C is a primary treatment target in the AACE and UMHS guidelines. Most of the other guidelines (SJTF, NHG, SIGN, ATPIII) don't use HDL-C as a treatment target. However, some of the guidelines (SJTF, ITF and ATPIII) pay special attention to patients with a specific dyslipidemia such as metabolic syndrome and isolated low HDL-C. According to the ATPIII, LDL-C remains the primary target in these patients. However, non-HDL-C is the secondary target of therapy. In patients whose TG are elevated along with LDL-C, it may be desirable to lower TG and increase HDL-C.

In the ITF guidelines, HDL-C is a primary target in patients with diabetes. The FMS guidelines suggest no target levels at all.

Management of Low HDL-C

Most of the guidelines don't focus on the therapeutic approach of low HDL-C or metabolic syndrome. For that reason the management of low HDL-C and therapy induced low HDL-C is mentioned only very briefly and incompletely. Most of the guidelines recommend higher doses of statins for low HDL-C during the treatment.

Only the ICSI guidelines focus on low HDL-C during the treatment with statins. They consider HDL-C as a secondary target after the LDL-C goal was achieved and they suggest fibrates or nicotinic acid for these patients.

Low HDL-C is, according to the SJTF, slightly increased by statins and resins and more substantially by nicotinic acids and fibrates.

The ATPIII guidelines propose nicotinic acid or fibrates for patients with isolated low HDL-C if CHD or risk equivalents exist.

Recommendations from the Guidelines

Although HDL-C levels below 40 mg/dl are considered as markers of an increased risk for CHD in most guidelines, HDL-C is not included into the coronary risk charts. These charts merely indicate that the CHD risk is higher for patients with low HDL-C. Unfortunately these guidelines don't take HDL-C into account for the risk calculation.

Most guidelines mention that there is insufficient evidence to justify goals for HDL-C. They claim that HDL-C should be used only to identify patients at higher risk. HDL-C below 40 mg/dl and a ratio of TC to HDL-C above 5 are considered as markers of higher risk. This was true at the time the guidelines were published because at that time the results of the VA-HIT were not available. The VA-HIT [20], the Helsinki Heart Study [7], the Münster Heart Study [32,48] and the BIP study [49] provide sufficient evidence to confirm the importance of HDL-C in cardiovascular prevention.

Most of the guidelines don't use the ratio of TC to HDL-C, although the Framingham function incorporates the ratio in the risk calculation. A lower ratio improves CHD risk prediction, particularly in women [23]. The Münster Heart Study also uses the ratio of TC to HDL-C in the risk calculation [32,48].

In the SJTF guidelines, HDL-C is not retained as a threshold or target for the treatment because HDL-C is not routinely measured across Europe. Therefore, the decision to use TC only was primarily made to ensure the widest possible application of the guidelines across Europe. A further consideration was that a chart based on the TC to HDL-C ratio would have to assume an average European cholesterol value. This was not appropriate for the European population because large differences in average cholesterol levels between countries exist [40].

The ratio of LDL-C to HDL-C is also a good indicator of risk. However, a mistake made in the measurement of HDL-C will affect the calculation of LDL-C with the Friedewald formula and compound the mistake in the assessment of risk. Very high levels of HDL-C will reduce the levels of TC calculated to be present in LDL-C and vice versa.

In conclusion, almost all guidelines consider low HDL-C as a marker of increased risk for CHD. However, only few of them use the HDL-C level as a threshold or target level for the treatment of dyslipidemia. Only the ICSI guideline provides any information on the management of patients with treatment-induced low HDL-C.

Instead of using TC or LDL-C, as a threshold as well as a target for the treatment, we consider the use of the ratio of TC to HDL-C or LDL-C to HDL-C. The use of the ratios provides much better discrimination of the patients at real risk and offers a better follow-up.

The Treatment of Bad HDL-C Responders

Low HDL-C can be secondary to other modifiable risk factors such as cigarette smoking, obesity, or physical inactivity. Therapeutic lifestyle changes that focus on these factors should be encouraged among the bad HDL-C responders. Especially weight reduction and physical activity should be emphasized.

In secondary prevention, the favorable results of the VA-HIT study have led some authorities to favor fibrates over statins in the treatment of patients with CHD and low-HDL. However, clinical trials with statins have been more robust in their outcome. In addition, the combined use of statins with either a fibrate or nicotinic acid is attractive for high-risk patients with isolated low HDL-C to improve the whole lipoprotein profile. However, using drugs in combination may increase the likelihood of side effects. Recent studies with ezetimibe, combined with statins, seem to be very promising to raise HDL-C without side effects [50].

In patients without CHD or CHD risk equivalents, low HDL-C counts as a risk factor that modifies the goal for LDL-C [1]. The first line of therapy for isolated low HDL-C is to maximize life habit changes. Whether fibrates or nicotinic acid may achieve a similar benefit in primary prevention as in secondary prevention is uncertain because primary prevention trials with these drugs have not targeted persons with isolated low HDL-C.

Several clinical trials suggest that raising HDL-C levels contributes to a decreased risk for CHD. For example, the AFCAPS/TexCAPS [51], the LCAS [52] and the LOCAT [53] provided information on the benefit of lipoprotein modification in patients with low baseline HDL-C. The VA-HIT study specially targeted patients with isolated low HDL-C for gemfibrozil therapy [20]. The reduction in major cardiovascular events was attributed in part to raising HDL-C. Likewise, the decrease in major coronary events during gemfibrozil therapy in the Helsinki Heart Study [5] was estimated to be partly due to an increase in HDL-C.

Nonetheless, in these trials, changes in other lipoproteins have also occurred. For this reason, the benefit of raising HDL-C is not recognized with certainty. On the other hand the conclusions of the landmark statins trials cannot be attributed to the decrease of LDL-C only because HDL-C also increased in these trials [7,8].

Especially for the 'bad HDL-C responders to statins' there is uncertainty about the pharmaceutical management of low HDL-C. Arguments exist to continue the statins, especially among patients with high LDL-C. However, we remarked that for some of our patients with low HDL-C during statin treatment, the HDL-C level was almost twice as high with a fibrate in monotherapy. No evidence about this was found in the literature. Switching these patients to fibrates in monotherapy or adding a fibrate to the statins could have a beneficial effect on the cardiovascular risk.

To confirm our preceeding findings, we analyzed all our 'bad HDL-C responders to statins' retrospectively. The aim was to describe the benefit of monotherapy with fibrates among these patients.

Data Collection

For this survey all patient records in our practice were analyzed over a 10-year period. All patients ever been treated with fibrates between 1991 and 2000 and having a decrease of HDL-C below 40 mg/dl during statin treatment were selected. In total 120 treated patients were screened. The inclusion criteria and the methodology of the measurements are already described in the third paragraph of this chapter.

No particular motives for selection of a statin or a fibrate at baseline existed. The personal and family medical history for major cardiac events, ischemic heart diseases, hypertension, stroke and diabetes were recorded cross-sectionally. Hypertension was diagnosed after at least two blood pressure measurements exceeding 160/100 mmHg. Diabetes was diagnosed for patients with fasting blood sugar exceeding 126 mg/dl.

SPSS-PC 10® (SPSS Inc.,Chicago,IL,USA) was used for analysis and statistical processing. Significant differences were detected with the independent-samples t-test. The cross-tables were used to detect differences between groups by means of chi-square tests.

Bad HDL-Responders

Fourteen 'bad HDL-C responders to statins' encountered the inclusion criteria. The mean age of the bad HDL-C responders was 58.6 years (SD 12.2) and ranged from 32 to 75 years. Eleven of the fourteen bad HDL-C responders were men. Their baseline lipid levels after a diet of at least three months are shown in table 6.

Their mean BMI was 26.8, 71% were treated for hypertension, 36% were current smokers, 36% had a history of myocardial infarction and 14% of stroke. Fourteen percent was diabetic. None of the patients had familial dyslipidemia.

The mean BMI of the 106 patients without bad HDL-C response to statins was 25.6, 58% were treated for hypertension, 28% were current smokers, 3% had a history of myocardial infarction and 3% of stroke. Ten percent was diabetic.

Crossover Results

Eight patients were first treated with fibrates and thereafter with statins and six received first statins and thereafter fibrates. Treatment with fibrates consisted of micronized fenofibrate 200 mg OD (n=8), ciprofibrate 100 mg OD (n=5) and bezafibrate 400 mg OD (n=1) and the treatment with statins consisted of pravastatin 20 mg OD (n=9), simvastatin 20 mg OD (n=4) and fluvastatin 40 mg OD (n=1). None of the patients was treated at the same time with statins and fibrates. The mean duration of the treatment was 1.2 year for the fibrates and 1.1 year for the statins. Table 6 shows the mean lipid levels at baseline and during statin and fibrate treatment.

Table 6. Crossover results between statins and fibrates for bad HDL-C responders to statins (n=14).

	Baseline	Statins	Fibrates	P
TC	276	226	243	> 0.05
LDL-C	192	157	167	> 0.05
HDL-C	44	33	49	< 0.001
TG	199	181	137	> 0.05
TC to HDL-C	6.5	7.0	5.2	< 0.01
LDL-C to HDL-C	4.5	4.9	3.6	< 0.001

TC and LDL-C were respectively 8% and 6% higher with fibrates compared to statins. TG were 24% lower with fibrates and HDL-C was 49% higher with fibrates compared to statins. HDL-C levels were higher for all patients during fibrate treatment. TC to HDL-C and LDL-C to HDL-C ratios were respectively 26% and 27% lower with fibrates. Changes in HDL-C and TC to HDL-C and LDL-C to HDL-C ratios were significant.

Choice between Statins and Fibrates?

The better results with the fibrates don't seem to be related to the younger age of the patients during the fibrate treatment. Six out of the fourteen patients received the fibrates after the statin. Bad HDL-C response to statins was not related to the type of statin used. Bad HDL-C response to statins was found among all statins. Comparison of the frequency of bad HDL-C responders to statins for the different statins is impossible with this dataset because the number of treated patients per group is too small. Bad HDL-C responders to atorvastatin were also found among our patients, but none of them ever received a fibrate. Therefore they were not included in this study. Simvastatin and pravastatin were prescribed only in low doses because they were launched on the Belgian market at doses of 20 mg. We may believe that the lipid levels may be better with higher doses of statins [54]. Although it was never tested among bad HDL-C responders to statins, we have serious doubt that the higher levels of HDL-C during fibrate treatment also could have been obtained with higher doses of statins.

The proportions of men, smokers and patients with hypertension, diabetes, myocardial infarction and stroke seems to be higher among the bad HDL-C responders than in our control population without bad HDL-C response to statins. Although it was not the primary objective of our study to evaluate cardiac events, it is noteworthy that the prevalence of serious cardiovascular events was higher among the bad HDL-C responders.

When switching the bad HDL-C responders from statins to fibrates, HDL-C increases significantly with more favorable TC to HDL-C and LDL to HDL-C ratios. The results of the switch of statin patients with low HDL-C to fibrates were never described in the medical literature. The switch from fibrates to statins is justifiable for patients with low or normal LDL-C. For patients with high LDL-C it might be better to add fibrates to the statins. The use of fibrates in patients with ischemic heart disease remains controversial because fibrates are known to increase homocysteine levels [55].

The genetic profile of the patient can also play an important role. The known polymorphisms in the beta-fibrinogen gene, the lipoprotein lipase gene and the hepatic lipase gene have an influence on the effects of statins in the general population [56]. The expectation is that in the future a subject's genotype may determine whether he will be treated with statins or fibrates.

Although HDL-C is taken into account in some cardiovascular risk calculation tables [57] it is not taken into account as target level for lipid-lowering treatment [1,40,46]. However, increased ratios of TC to HDL-C and LDL-C to HDL-C as a result of decreased HDL-C levels are quite uncomfortable because a higher cardiovascular risk score could be obtained during the treatment with the statin than before the treatment.

The importance of low HDL-C on coronary endpoints, even in the presence of normal TC and LDL-C levels has been shown in the VA-HIT study [20]. It was the first double-blind trial demonstrating that raising the HDL-C with fibrates resulted in a significant reduction in the risk of major cardiovascular events in patients with CHD whose primary lipid abnormality was a low HDL-C level [20]. However, it should be mentioned that only 25% of the risk reduction in the VA-HIT study was attributed to changes in HDL-C. Other factors such as peroxisome proliferator activated receptor (PPAR) activity may also contribute to the risk reduction. Statins may also have some pleotrophic effects which could contribute to risk reduction and they may not be related to changes in HDL-C.

Besides the pharmaceutical treatment, the patients with low HDL-C should receive an appropriate advice concerning therapeutic lifestyle changes such as weight reduction, reduced intake of cholesterol-raising nutrients and increased physical activity.

Review and Critical Appraisal of the RCTs with Fibrates

In the Helsinki Heart Study 4081 asymptomatic middle-aged men (40 to 55 years of age) with primary dyslipidemia (non-HDL cholesterol > or = 200 mg/dl) were randomized to gemfibrozil or placebo [5]. After five years the incidence of fatal or nonfatal coronary heart disease was reduced with 34% in the gemfibrozil group ($p<0.02$). There was no difference between the groups in the total death rate, nor did the treatment influence the cancer rates.

The Lopid Coronary Angiographic Trial (LOCAT) was the first landmark trial with gemfibrozil [58]. In total 372 men who had undergone coronary artery bypass (CABG) and had low HDL-C (< or = 42.5 mg/dl) were randomized to placebo or gemfibrozil. Serum TG was < or = 354 mg/dl and LDL-C was < or = 174 mg/dl. After a follow-up of two years 372 finished the study with suitable angiograms. The triglyceride rich lipoproteins, especially Intermediate-density lipoprotein cholesterol (IDL-C), were predictors of angiographic progression, as well as elevated LDL-C. Of the HDL-C subfractions only HDL_3-C was protective.

In a sub-analysis of the Bezafibrate Coronary Atherosclerosis Intervention Trial (BECAIT) 81 male survivors of a first myocardial infarction under 45 years of age with TC of at least 200 mg/dl and TG of at least 142 mg/dl were randomized to bezafibrate or placebo

[59]. After 5 years bezafibrate retarded the progression of focal coronary atherosclerosis and this effect could be attributed at least in part to an increase in HDL_3-C and a decrease in apo B containing lipoproteins.

In the Veterans Affairs High-Density Lipoprotein Cholesterol Intervention Trial (VA-HIT) 2531 men with coronary artery disease (history of myocardial infarction, angina with objective evidence of ischemia, coronary revascularization, coronary stenosis >50% by angiography) were randomized to gemfibrozil or placebo [20]. They had low HDL-C (< or = 40 mg/dl) and low LDL-C (< or = 140 mg/dl). In men with coronary artery disease, whose main lipid abnormality was low HDL-C, gemfibrozil increased HDL-C, lowered TG and reduced the risk of major cardiovascular events. The relative risk for the primary endpoint (nonfatal myocardial infarction or death from coronary artery disease) decreased with 22% (p=0.006).

HDL-C levels during treatment with gemfibrozil predicted the magnitude of reduction in risk for coronary heart disease events in patients with low HDL-C at baseline. However changes in HDL-C concentrations only partly explained the reduction in event rates by gemfibrozil [60].

In the VA-HIT Stroke Substudy gemfibrozil significantly reduced the incidence of strokes in men with CAD, low HDL-C and low LDL-C. The relative risk for stroke decreased with 31% (p=0.036) [61].

In the Bezafibrate Infarction Prevention (BIP) 3090 patients with previous myocardial infarction or unstable angina were randomized to bezafibrate or placebo [49]. HDL-C was < or = 45 mg/dl, TG < or = 300 mg/dl, TC between 180 and 250 mg/dl and LDL-C < or = 180 mg/dl. After 6.2 years bezafibrate had not significantly decrease the primary cardiac event rate. In a post hoc subgroup analysis it did decrease the incidence of fatal or nonfatal myocardial infarctions or sudden death in coronary patients with a high baseline TG (> or = 200 mg/dl).

In the Diabetes Atherosclerosis Intervention Study (DAIS) 731 men and women with type 2 diabetes were randomly assigned to fenofibrate or placebo for at least 3 years [62]. They were in good glycaemic control and had mild lipoprotein abnormalities. Half had at least one visible coronary lesion and half had no previous clinical coronary disease. The fenofibrate group showed a significantly smaller increase in percentage diameter stenosis than the placebo group (p=0.02), a significantly smaller decrease in minimum lumen diameter (p=0.029), and a non-significantly smaller decrease in mean segment diameter (p=0.171). The trial was not powered to examine clinical endpoints, but there were fewer events in the fenofibrate group than in the placebo group (38 vs 50). The DAIS suggests that treatment with fenofibrate reduces the angiographic progression of coronary-artery disease in type 2 diabetes. This effect is related, at least partly, to the correction of lipoprotein abnormalities, even those previously judged not to need treatment.

In conclusion, there is convincing epidemiological evidence for the role of low HDL-C in CHD risk. However the RCTs are less convincing. The Helsinki Heart Study, the LOCAT study and the VA-HIT study are studies with gemfibrozil, which is no longer available in Belgium. The BECAIT and the BIP study use bezafibrate. The DIAS is the only important study with fenofibrate. In most of these studies selected populations are being studied and different endpoints are taken into account. The studies on surrogate endpoints such as angiographic parameters are partly convincing. The statistical and clinical significance of increasing HDL-C on hard endpoints is often very low or absent.

Treatment of Low HDL-C

The first step of therapy for isolated low HDL-C is to maximize life habit changes. These include all components of therapeutic lifestyle changes (reduction in cholesterol-raising nutrients, weight reduction and increased physical activity).

In patients with CHD or risk equivalents, the primary target for therapy is LDL-C. Once adequate LDL-C levels have been obtained, other lipid risk factors deserve attention. Nicotinic acid and fibrates can be considered if the patient has low HDL-C.

Fibrates usually raise HDL-C by 10-15% but greater increases can occur in persons with very high TG and very low HDL-C levels [1]. Micronised fenofibrate has shown to significantly increase mean HDL-C levels to above 40 mg/dl, irrespective of baseline HDL-C levels [19].

Nicotinic acid can be used instead of a fibrate. It has the advantage of raising HDL-C two-to-threefold more times than fibrates [63]. With statins HDL-C generally rises by 5-10% but greater increases usually occur in patients with low HDL-C and elevated TG [24].

Combined drug therapy (low-dose statins and fibrates or nicotinic acids) remains an option for patients with low HDL-C, provided that precautions are taken to prevent and monitor the side effects of lipid-lowering drugs used in combination. A major safety concern with combined drug therapy is the rare occurrence of myopathy and rhabdomyolisis [64]. Recent studies with ezetimibe, combined with statins, seem to be very promising to raise HDL-C without side effects [65].

The switch of patients with low HDL-C during statin treatment to a treatment with fibrates can result in increased levels of HDL-C with more favorable ratios of TC to HDL-C and LDL-C to HDL-C [14].

In conclusion, this study addresses the value of fibrates to increase HDL-C levels to achieve the 40mg/dl level in those who might fail to achieve this with statins. Patients with low HDL-C during the treatment with statins had far better HDL, TC to HDL-C and LDL-C to HDL-C after they had been switched to fibrates. Compared to the control group these patients, whom we called 'bad HDL-C responders to statins', were characterized by an increased prevalence of myocardial infarction.

The Future Research on Bad HDL-C Responders to Statins

As a result of the research on bad HDL-C responders to statins, a RCT was designed to prove that fibrates are preferable over statins in the treatment of dyslipidaemia among subjects with a bad HDL-C response to statins. For this purpose a cross-over trial was designed including 400 patients presenting low HDL-C during statin treatment (Figure 2). Two hundred patients had to be randomized to simvastatin treatment and two hundred other patients to fenofibrate. After 4 months of treatment the patients receiving simvastatin should have been switched to fenofibrate and the patients initially receiving fenofibrate should have been switched to simvastatin. This cross-over was needed to provide enough power to the study because difficulties could be expected to include sufficient patients to demonstrate a significant difference between both groups. The cross-over was also a method to avoid study bias caused by differences between both study groups.

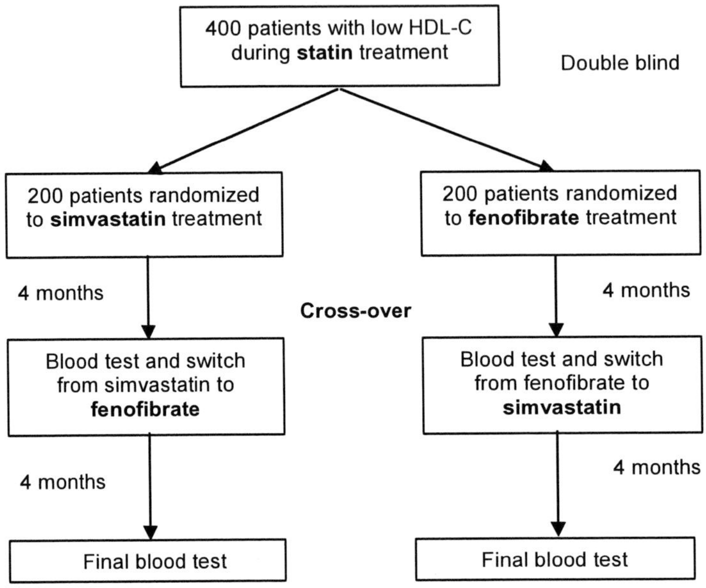

Figure 2. Study design of the RCT for "bad HDL-C responders to statins"

The RCT was not accomplished because there were insurmountable problems to find the necessary funds. The role of the pharmaceutical companies in this matter is not to be underestimated. Companies are more interested in promoting expensive drugs such as statins instead of inexpensive drugs such as fibrates. They are prepared to invest huge amounts of money to sponsor RCTs that are not initiated and performed by their companies, because they know very well that RCTs, which are performed by the companies themselves, have a lower credibility than RCTs, which are performed by independent research institutes. However pharmaceutical companies sponsor almost all these research institutes. Pharmaceutical companies are also prepared to invest a lot of manpower in the design of RCTs. The design of the most recent RCTs in terms of set-up, inclusion and exclusion criteria and study power are examples of high tech. These studies have a well-considered design, which permits with an almost one hundred percent certainty the confirmation of the aims of the study that are predefined by the companies. The studies which are initiated to prove the beneficial effect of the molecule are often more expensive than the development of the molecule itself.

This is an example of the priorities given by pharmaceutical companies for their research. They are not prepared to invest money in drugs which are not expensive or for which the patent is about to expire. The implementation of a RCT without the sponsoring from pharmaceutical companies is almost impossible.

Conclusion

In this chapter the hypothesis about "bad HDL-C responders to statins" was formulated. A far better lipoprotein profile was achieved with fenofibrate in the patient with low HDL-C during pravastatin treatment. These findings were confirmed by an observational study including all patients of our practice treated with lipid-lowering drugs.

Practice observation confirmed the value of fibrates, which were prescribed to increase HDL-C levels, in trying to reach the target 40 mg/dl level in patients who failed to achieve this goal with statins. The concerned patients were characterized by low baseline HDL-C, male gender and an increased risk of myocardial infarction.

The next step in our study about bad HDL-C responders to statins was a review of the lipid-lowering guidelines. Almost all guidelines consider low HDL-C as a marker of increased risk for CHD. However only few of them use the HDL-C level as a threshold or target level for the treatment of dyslipidaemia. The VA-HIT trial indicates that there is now evidence for making HDL-C just as prominent a target as LDL-C. Instead of using TC or LDL-C as a threshold as well as a target for the treatment, one should consider the use of the ratio of TC to HDL-C or LDL-C to HDL-C.

The Impact of the Research

The research on bad HDL-C responders may have an impact on the clinical guidelines. In the review suggestions were made for an adaptation of the guidelines according to our findings and according to the more convincing results of the VA-HIT. We are aware of the fact that the level of evidence of the VA-HIT is much higher than for our observational studies and review.

The above-mentioned research will certainly have an impact on future research. We didn't succeed to implement a RCT in which fibrates and statins are compared for bad HDL-C responders. The European branch of Fournier Pharma, however, initiated a RCT comparing fenofibrate and simvastatin among patients with low baseline HDL-C.

The research could also inspire basic scientists to investigate the biological base for bad HDL-C response to statins.

Acknowledgements

We thank the co-authors of the papers: Brigitte Velkeniers, William Duquet, Wim Betz, Jan Vandevoorde, Jan Kartounian, Karolien Vantomme, Rinaldo Lauwers and Peter Coigniez.

The authors thank Jean-Pierre Bronckaers, Cary Goovaerts, Jef Bresseleers and Jacques Vergucht of the Liberale Mutualiteit Brabant, Nick De Swaef and Martine Duyck of the Vlaams Neutraal Ziekenfonds Aalst, Vanessa Meert for the data-entry and Michael Egan, Marianne van Winden and Neeltje Blommaert for their comments.

REFERENCES

[1] Executive Summary of The Third Report of The National Cholesterol Education Program (NCEP) Expert Panel on Detection, Evaluation, And Treatment of High Blood Cholesterol In Adults (Adult Treatment Panel III). *JAMA* 2001;285:2486-97.

[2] Shepherd J, Cobbe SM, Ford I, et al. Prevention of coronary heart disease with pravastatin in men with hypercholesterolemia. *N Engl J Med* 1995;333:1301-7.

[3] Pedersen TR, Olsson AG, Faergeman O, et al. Lipoprotein changes and reduction in the incidence of major coronary heart disease events in the Scandinavian Simvastatin Survival Study (4S). *Circulation* 1998;97:1453-60.

[4] Sacks FM, Moyé LA, Davis BR, et al. Relationship between plasma LDL concentrations during treatment with pravastatin and recurrent coronary events in the cholesterol and recurrent events trial. *Circulation* 1998;97:1446-52.

[5] Frick MH, Elo O, Haapa K, et al. Helsinki Heart Study: primary-prevention trial with gemfibrozil in middle-aged men with dyslipidaemia. Safety of treatment, changes in risk factors, and incidence of coronary heart disease. *N Engl J Med*, 1987;317:1237-45.

[6] Wilson PW, Abbott RD, Castelli WP. High-density lipoprotein cholesterol and mortality. The Framingham Heart Study. *Arteriosclerosis* 1988;8:737-41.

[7] Sacks FM, Pfeffer MA, Moye LA, et al. The effect of pravastatin on coronary events after myocardial infarction in patients with average cholesterol levels. *N Engl J Med* 1996;335:1001-9.

[8] Scandinavian Simvastatin Survival Study Group. Randomised trial of cholesterol lowering in 4444 patients with coronary heart disease: the Scandinavian Simvastatin Survival Study (4S). *Lancet* 1994;344:1383-9.

[9] Barr DP, Russ EM, Eder HA. Protein-lipid relationships in human plasma. *Am J Med* 1951;11:480-5.

[10] Castelli WP, Garrison RJ, Wilson PW, Abbott RD, Kalousdian S, Kannel WB. Incidence of coronary heart disease and lipoprotein cholesterol levels. The Framingham Study. *JAMA* 1986;256:2835-8.

[11] Rubins HB, Robins SJ, Collins D, et al. Distribution of lipids in 8,500 men with coronary artery disease. *Am J Cardiol* 1995;75:1196-201.

[12] Gotto AM Jr. High-density lipoprotein cholesterol and triglycerides as therapeutic targets for preventing and treating coronary artery disease. *Am Heart J* 2002;144:S33-42.

[13] Prevention of Cardiovascular Events and Death with Pravastatin in Patients with Coronary Heart Disease and a Broad Range of Initial Cholesterol Levels. The Long-Term Intervention with Pravastatin in Ischaemic Disease (LIPID) Study Group. *N Engl J Med* 1998;339:1349-57.

[14] Devroey D, Betz W, Coigniez P, et al. A 'bad responder' to statins. *Cardiology* 2002;97:230-2.

[15] Devroey D, Velkeniers B, Duquet W, Betz W. Serum lipid comparison in patients treated by statins or fibrates: existence of bad HDL-C responders to statins. *Acta Cardiol* 2003;58:179-184.

[16] Devroey D, Velkeniers B, Betz W, Kartounian J. The prevalence of low levels of high-density lipoprotein cholesterol among patients treated with lipid-lowering drugs. *Brit J Cardiol* 2004;11:50-55.

[17] Devroey D, Vantomme K, Betz W, Vandevoorde J, Kartounian J. A review of the treatment guidelines on the management of low levels of high-density lipoprotein cholesterol. *Cardiology* 2004;102:61-66.

[18] Devroey D, Velkeniers B, Duquet W, Betz W. The benefit of fibrates in the treatment of 'bad HDL-C responders to statins'. *Int J Cardiol* (in press).

[19] Poulter N. The impact of micronized fenofibrate on lipid subfractions and on reaching HDL-target levels in 7,098 patients with dyslipidaemia. *Br J Card* 1999;6:682-5.

[20] Rubins HB, Robins SJ, Collins D, et al. Gemfibrozil for the secondary prevention of coronary heart disease in men with low levels of high-density lipoprotein cholesterol. Veterans Affairs High-Density Lipoprotein Cholesterol Intervention Trial Study Group. *N Engl J Med* 1999;341:410-8.

[21] Krauss RM, Deckelbaum RJ, Ernst N, et al. Dietary guidelines for healthy American adults: a statement for health professionals from the Nutrition Committee, American Heart Association. *Circulation* 1996;94:1795–800.

[22] Friedewald WT, Levy RI, Fredrickson DS. Estimation of the concentration of low-density lipoprotein cholesterol in plasma, without use of the preparative ultracentrifuge. *Clin Chem* 1972;18:499-502.

[23] Wilson PW, D'Agostino RB, Levy D, Belanger AM, Silbershatz H, Kannel WB. Prediction of coronary heart disease using risk factor categories. *Circulation* 1998;97:1837-47.

[24] LaRosa JC, He J, Vupputuri S. Effect of statins on risk of coronary disease: a meta-analysis of randomized controlled trials. *JAMA* 1999;282:2340-6.

[25] Leaf DA, Connor WE, Illingworth DR, Bacon SP, Sexton G. The hypolipidemic effects of gemfibrozil in type V hyperlipidemia. A double-blind, crossover study. *JAMA* 1989;262:3154-60.

[26] Muls E, De Backer G, Brohet C, Heller F. The efficacy of atorvastatin in treating patients with hypercholesterolaemia to target LDL-cholesterol goals: the LIPI-GOAL trial. *Acta Cardiol* 2001;56:109-14.

[27] Svilaas A, Thoresen M, Kristoffersen JE, Hjartaaker J, Westheim A. How well are patients with atherosclerotic disease treated? Secondary prevention in primary care. *Scand J Prim Health Care* 2000;18:232-6.

[28] Vanuzzo D, Pilotto L, Ambrosio GB, Pyorala K, Lehto S, De Bacquer D, De Backer G, Wood D. Potential for cholesterol lowering in secondary prevention of coronary heart disease in europe: findings from EUROASPIRE study. European Action on Secondary Prevention through Intervention to Reduce Events. *Atherosclerosis* 2000;153:505-17.

[29] Gordon T, Castelli WP, Hjortland MC, Kannel WB, Dawber TR. High density lipoprotein as a protective factor against coronary heart disease. The Framingham Study. *Am J Med* 1977;62:707-714.

[30] Assmann G, Cullen P. Schulte H. The Münster Heart Study (PROCAM). Results of follow-up at 8 years. *Eur Heart J* 1998;19:A2-A11.

[31] Gordon DJ, Probstfield JL, Garrison RJ, Neaton JD, Castelli WP, Knoke JD, Jacobs DR Jr, Bangdiwala S, Tyroler HA. High-density lipoprotein cholesterol and cardiovascular disease. Four prospective American studies. *Circulation* 1989;79:8-15.

[32] Knuiman JT, Westenbrink S, van der Heyden L, et al. Determinants of total and high density lipoprotein cholesterol in boys from Finland, The Netherlands, Italy, the Philippines and Ghana with special reference to diet. *Hum Nutr Clin Nutr* 1983;37:237-54.

[33] Maitland-van der Zee AH, Klungel OH, Stricker BH, et al. Genetic polymorphisms: importance for response to HMG-CoA reductase inhibitors. *Atherosclerosis* 2002;163:213-22.

[34] O'Neill FH, Patel DD, Knight BL, et al. Determinants of variable response to statin treatment in patients with refractory familial hypercholesterolemia. *Arterioscler Thromb Vasc Biol* 2001;21:832-7.

[35] Mahley RW, Palaoglu KE, Atak Z, et al. Turkish Heart Study: Lipids, lipoproteins, and apolipoproteins. *J Lipid Res* 1995;36:839-859.

[36] Bersot TP, Vega GL, Grundy SM, et al. Elevated hepatic lipase activity and low levels of high density lipoprotein in a normotriglyceridemic, nonobese Turkish population. *J Lipid Res* 1999;40:432-38.

[37] Cluzeau FA, Littlejohns P, Grimshaw JM, et al. Development and application of a generic methodology to assess the quality of clinical guidelines. *Int J Qual Health Care* 1999;11:21-8.

[38] Burgers JS, Cluzeau FA, Hanna SE, et al. Characteristics of high quality guidelines: evaluation of 86 clinical guidelines developed in ten European countries and Canada. *Int J of Technology Assessment in Health Care* 2002;18:955-964.

[39] Burgers JS, Bailey JV, Klazinga NS, et al. Inside guidelines: comparative analysis of recommendations and evidence in diabetes guidelines from 13 countries. *Diabetes Care* 2002;25:1933-1939.

[40] Wood D, De Backer G, Faergeman O, et al. Prevention of coronary heart disease in clinical practice: recommendations of the Second Joint Task Force of European and other Societies on Coronary Prevention. *Atherosclerosis* 1998;140:199-270.

[41] Thomas S, Van der Weijden T, Van Drenth B, et al. NHG-Standaard Cholesterol. *Huis Wet* 1999;42:406-417.

[42] Finnish Medical Society Duodecim. Drug treatment for hyperlipidaemias. In: EBM Guidelines Evidence-Based Medicine (CDROM). Helsinki (Finland): *Duodecim Medical Publications,* Ltd; 2001.

[43] Scottish Intercollegiate guidelines Network (SIGN). Secondary Prevention of Coronarty Heart Disease following Myocardial Infarction. *A National Clinical Guideline*; 2000.

[44] American Association of Clinical Endocrinologists (AACE) medical guidelines for clinical practice for the diagnosis and treatment of dyslipidemia and prevention of atherogenesis. *Endocr Pract* 2000;6:162-213.

[45] University of Michigan Health System (UMHS). Screening and Management of Lipids. In: *Guidelines for Clinical care*. University of Michigan; 2000.

[46] Institute for Clinical Systems improvement (ICSI). Treatment of lipid disorders in adults. In: *ICSI Health Care Guidelines*. Bloomington; 2002.

[47] The International Task Force for Prevention of Coronary Heart Disease. The scientific background for primary and secondary prevention of coronary heart disease. *Nutr Metab Cardiovasc Dis* 1998;8:205-271.

[48] Assmann G, Schulte H, von Eckardstein A, et al. High-density lipoprotein cholesterol as a predictor of coronary heart disease risk. The PROCAM experience and pathophysiological implications for reverse cholesterol transport. *Atherosclerosis* 1996;124:S11-S20.

[49] The BIP Study Group. Secondary prevention by raising HDL-cholesterol and reducing triglycerides in patients with coronary artery disease: The Bezatibrate Infarction Prevention (BIP) Study. *Circulation* 2000;102:21-7.

[50] Gagne C, Bays HE, Weiss SR, et al. Efficacy and safety of ezetimibe added to ongoing statin therapy for treatment of patients with primary hypercholesterolemia. *Am J Cardiol* 2002;90:1084-91.

[51] Downs GR. Clearfield M, Weiss S, et al. Primary prevention of acute coronary events with lovastatin in men and women with average cholesterol levels: results of AFCAPS/TEXCAPS (Air Force/Texas Coronary Atherosclerosis Study). *JAMA* 1998;279:1615-22.

[52] Herd JA, Ballantyne CM, Farmer JA, et al. Effects of fiuvastatin on coronary atherosclerosis in patients with mild to moderate cholesterol elevations (Lipoprotein and Coronary Atherosclerosis Study, LCAS). *Am J Cardiol* 1997;80:278-86.

[53] Frick MH, Syvanne M, Nieminen MS, et al. Prevention of the angiographic progression of coronary and vein-graft atherosclerosis by gemfibrozil after coronary bypass surgery in men with low levels of HDL-cholesterol. Lopid Coronary Angiography Trial (LOCAT) Study Group. *Circulation* 1997;96:2137-43.

[54] Jones PH, Davidson MH, Stein EA, et al. Comparison of the efficacy and safety of rosuvastatin versus atorvastatin, simvastatin, and pravastatin across doses (STELLAR Trial). *Am J Cardiol* 2003;92:152-160.

[55] Milionis HJ, Papakostas J, Kakafika A, et al. Comparative effects of atorvastatin, simvastatin, and fenofibrate on serum homocysteine levels in patients with primary hyperlipidemia. *J Clin Pharmacol* 2003;43:825-830.

[56] O'Neill FH, Patel DD, Knight BL, et al. Determinants of variable response to statin treatment in patients with refractory familial hypercholesterolemia. *Arterioscler Thromb Vasc Biol* 2001;21:832-837.

[57] Conroy RM, Pyörälä K, Fitzgerald AP, et al, on behalf of the SCORE project group. Estimation of ten-year risk of fatal cardiovascular disease in Europe: the SCORE project. *Eur Heart J*, 2003;24:987-1003.

[58] Syvanne, M., Nieminen, M. S., Frick, M. H., Kauma, H., Majahalme, S., Virtanen, V., Kesaniemi, Y. A., Pasternack, A., Ehnholm, C., & Taskinen, M. R. 1998, "Associations between lipoproteins and the progression of coronary and vein-graft atherosclerosis in a controlled trial with gemfibrozil in men with low baseline levels of HDL cholesterol", *Circulation*, vol. 98, no. 19, pp. 1993-1999.

[59] Ruotolo G, Ericsson CG, Tettamanti C, Karpe F, Grip L, Svane B, Nilsson J, de Faire U, Hamsten A. 1998, "Treatment effects on serum lipoprotein lipids, apolipoproteins and low density lipoprotein particle size and relationships of lipoprotein variables to progression of coronary artery disease in the Bezafibrate Coronary Atherosclerosis Intervention Trial (BECAIT)", *J Am Coll.Cardiol*, vol. 32, no. 6, pp. 1648-1656.

[60] Robins SJ, Collins D, Wittes JT, Papademetriou V, Deedwania PC, Schaefer EJ, McNamara JR, Kashyap ML, Hershman JM, Wexler LF, Rubins HB; VA-HIT Study Group. Veterans Affairs High-Density Lipoprotein Intervention Trial. Relation of gemfibrozil treatment and lipid levels with major coronary events: VA-HIT: a randomized controlled trial. *JAMA* 2001;285:1585-1591.

[61] Bloomfield RH, Davenport J, Babikian V, Brass LM, Collins D, Wexler L, Wagner S, Papademetriou V, Rutan G, Robins SJ. Reduction in stroke with gemfibrozil in men with coronary heart disease and low HDL cholesterol: The Veterans Affairs HDL Intervention Trial (VA-HIT). *Circulation* 2001;103:2828-2833.

[62] Effect of fenofibrate on progression of coronary-artery disease in type 2 diabetes: the Diabetes Atherosclerosis Intervention Study, a randomised study. *Lancet* 2001;357:905-910.

[63] Vega GL, Grundy SM. Lipoprotein responses to treatment with lovastatin, gemfibrozil, and nicotinic acid in normolipidemic patients with hypoalphalipoproteinemia. *Arch Intern Med* 1994;154:73-82.

[64] Pasternak RC, Smith SC Jr, Bairey-Merz CN, et al. ACC/AHA/NHLBI Clinical Advisory on the Use and Safety of Statins. *Circulation* 2002;106:1024-8.

[65] Gagne C, Bays HE, Weiss SR, et al. Efficacy and safety of ezetimibe added to ongoing statin therapy for treatment of patients with primary hypercholesterolemia. *Am J Cardiol* 2002;90:1084-91.

In: Statins: Indications and Uses, Safety and Modes of Action ISBN 978-1-60692-103-6
Editor: Gunnar N. Holmquist

Chapter 8

BENEFICIAL EFFECTS OF STATINS IN SYSTEMIC LUPUS ERYTHEMATOSUS: MOLECULAR MECHANISM INVOLVED

Antonio G. Tristano*

Department of Pharmaceutical and Administrative Sciences, College of Pharmacy, Health Professions Division, Nova Southeastern University (NSU), Fort Lauderdale, USA.

ABSTRACT

The 3-hydroxy-3-methylglutaryl-coenzyme-A (HMG-CoA)-reductase inhibitors (statins) are the most commonly prescribed agents for the treatment of hypercholesterolemia, due to their efficacy in lowering LDL-cholesterol and ability to reduce clinical outcome in both primary and secondary prevention of coronary artery disease. In addition to their serum lipid-lowering action, statins display non lipid-lowering pharmacological activities known as pleotropic actions. The pleiotropic effects include significant anti-inflammatory and immunomodulatory actions and many essential cellular functions including cell proliferation, differentiation, and survival and participate in the regulation of cell shape and motility.

This chapter is about the intracellular pathways involved in the pathogenesis of systemic lupus erithematosus (SLE) and the possible effects of the statins in those pathways which could be modulating their therapeutic effects, including their beneficial effects on primary and secondary prevention of cardiovascular diseases, anti-inflammation, and immunomululation.

* Correspondence concerning this article should be addressed to: Antonio G Tristano, 5150 NW 99th Ave. Miami, FL 33178. USA. E-mail: mjtristano@cantv.net.

Introduction

HMG-CoA reductase catalyzes the conversion of HMG-CoA to mevalonic acid, representing the rate limiting step in biosynthesis of cholesterol, as well as in isoprenoid intermediates such as farnesyl pyrophosphate (FPP) and geranylgeranyl pyrophosphate (GGPP), some of which become post-translationally incorporated into specific proteins [1-4]. Primarily, these proteins include the Ras, Rho and Rab family GTPases, which are responsible for cell growth, survival, and differentiation [5-7]. These GTPases link the extracellular stimuli to signaling molecules such as mitogenactivated protein kinases (MAPKs). MAPKs constitute a superfamily of serine/threonine proteins kinases involved in the regulation of a numbers of intracellular pathways. Mammalian MAPKs are grouped into five major subfamilies and have been associated with cellular growth and apoptosis, cellular differentiation and transformation, and vascular contraction [8-17].Of these, the ERK 1/2 (ERK 1 and ERK 2) are activated in response to growth and differentiation factors, while the JNKs and p38 MAPK are usually activated in response to inflammatory cytokines, cellular stresses, and withdrawal of growth factors [8,16,18-21].Therefore, the noncholesterol or pleiotropic effects of statins involve interrupting the composition of cell membranes and inhibiting protein prenylation. The pleiotropic effects include significant anti-inflammatory and immunomodulatory actions [22-30]. In part, the anti-inflammatory effects of statins can result from their capacity to interfere with the mevalonate pathway and inhibit prenylation of Rho family GTPases and MAPKs phosphorylation [31-36].

Systemic lupus erythematosus (SLE) is an inflammatory rheumatic disease of immunologic origin. SLE patients have an increased risk of myocardial infarction, and prevalence of atheroma, in part associated to the atherogenic lipid abnormalities seen in these patients, but also due to the role played by proinflammatory cytokines such as TNF-α, IL-1β, IL-6 [37-43]. Therefore, statins could have a dual benefit in SLE patients, either in the inflammatory process by itself or in the prevention of cardiovascular disease (CVD). This chapter will provide new insights for the use of statins in treating systemic SLE both for their anti-atherosclerotic activity and for their pleiotropic effects on inflammation, and the immune responses.

Mechanism of Action of Statins

Statins or HMG-CoA reductase inhibitors, are competitive inhibitors of the rate limiting enzyme in cholesterol biosynthesis and are usually prescribed for the treatment of hypercholesterolemia, due to their efficacy in lowering LDL and their ability to improve clinical endpoints in both primary and secondary prevention of coronary artery disease [1,44-49]. The chemical structures of the more frequent prescribed statins are shown in Figure 1.

HMG-CoA reductase catalyzes the conversion of HMG-CoA to mevalonic acid, representing the rate-limiting step in biosynthesis of cholesterol, as well as in isoprenoid intermediates such as farnesyl pyrophosphate (FPP) and geranylgeranyl pyrophosphate (GGPP) [50,51], some of which become post-translationally incorporated into specific proteins (Figure 2) [1,4]. Primarily, these proteins include the Ras, Rho and Rab family

GTPases, which are the responsible for cell growth, survival, and differentiation [5-7]. These GTPases link the extracellular stimuli to signaling molecules such as MAPKs. Therefore, the non-cholesterol or pleiotropic effects of statins involve interrupting the composition of cell membranes and inhibiting protein prenylation (Figure 3).

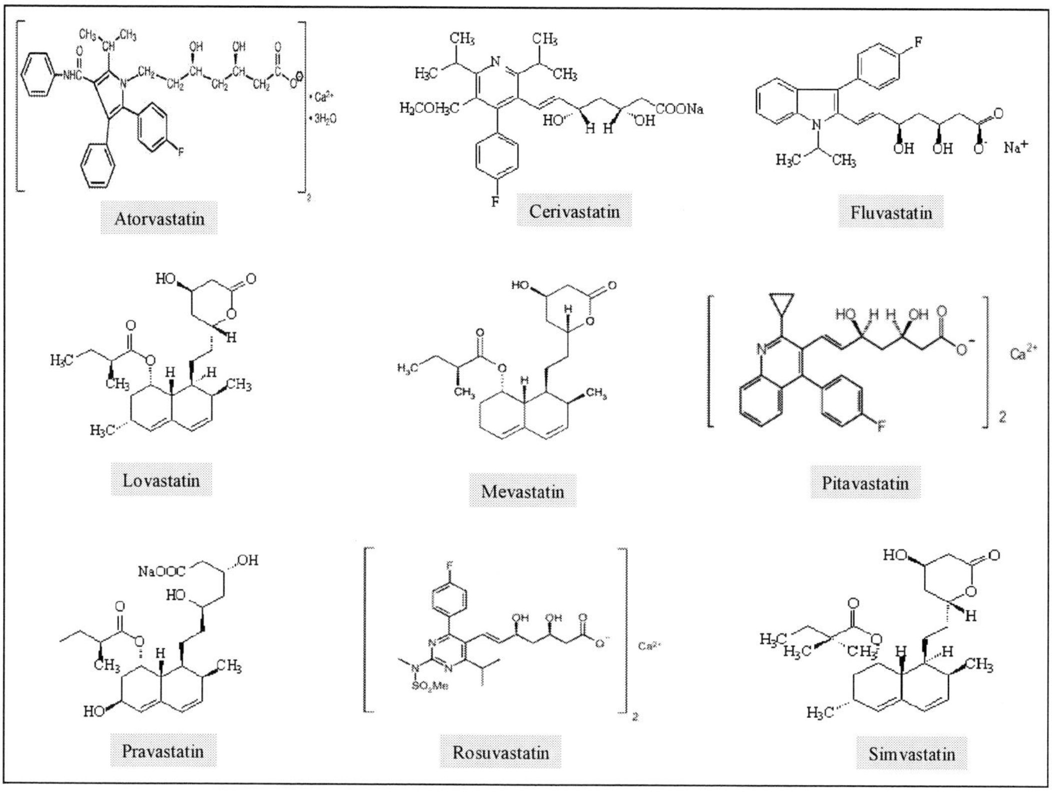

Figure 1. Chemical structures of the more frequent prescribed statins.

MITOGEN-ACTIVATED PROTEIN KINASES (MAPKS) PATHWAYS

MAPKs, are found in all eukaryotes and evolutionarily well conserved in cells [18,52]. MAPKs are a family of serine/threonine protein kinases that mediate nuclear transduction of extracellular signals by intracellular proteins phosphorylation, leading to a cascade of transcription factor activation, enhanced gene expression and throphic cellular responses [53,54]. Mammalian MAPKs are grouped into five major subfamilies and have been associated with cellular growth and apoptosis, cellular differentiation and transformation, and vascular contraction [8-17].

The MAP kinase-signalling cascade is typically composed of three protein kinases including MAP kinase (MAPK), MAPK kinase (MAPKK) and MAPKK kinase (MAPKKK). [55-60]. MAPKKK phosphorylates and activates a dual-specificity protein kinase (MAPKK), which in turn phosphorylates and activates another protein kinase (MAPK) [8,61]. Activated

MAPK then regulates the activities of transcription factors or kinases further downstream by phosphorylation, and thereby controls gene expression and cellular function.

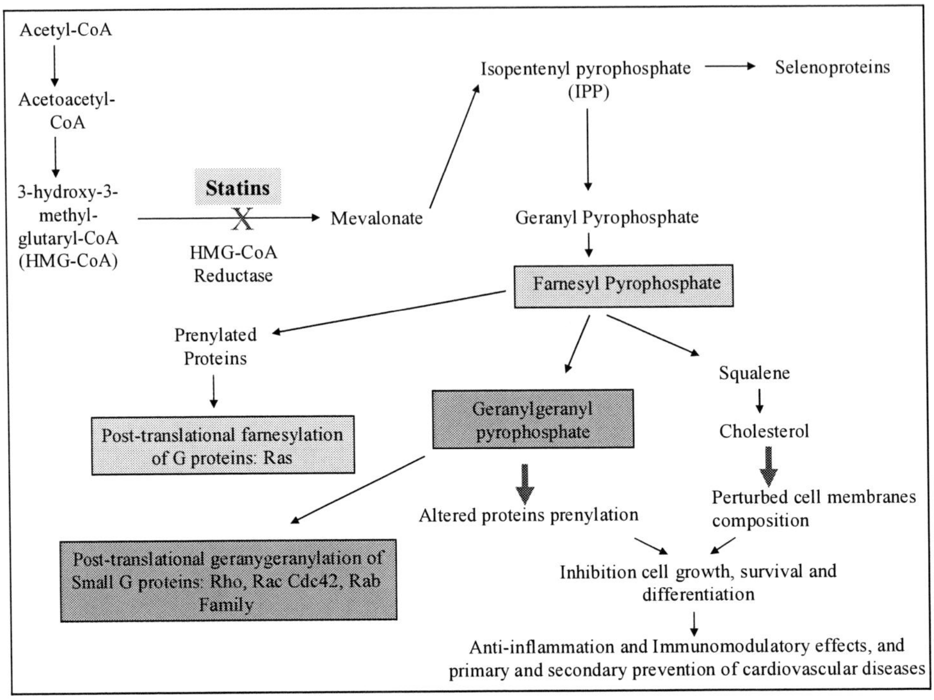

Figure 2. Effects of statins on protein prenylation, and cell membranes composition.

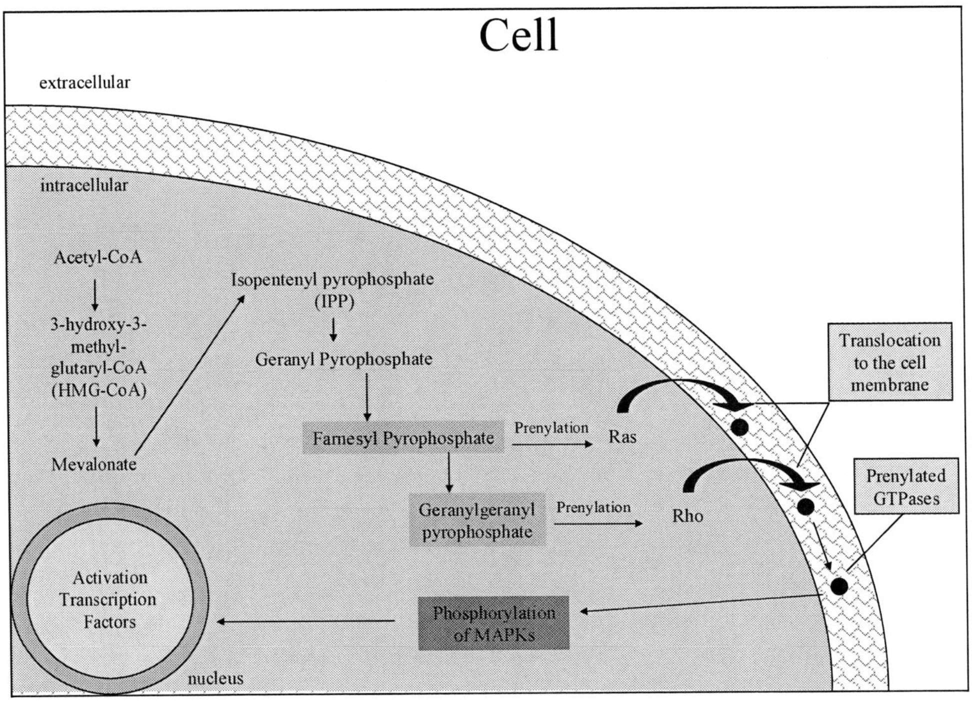

Figure 3. The mevalonate biosynthetic pathway and MAPKs phosphorylation.

MAPKs constitute a superfamily of serine/threonine proteins kinases involved in the regulation of a numbers of intracellular pathways (Figure 4). The mammalian MAP kinase family includes: a) ERK 1 and ERK 2, often referred as p44 and p42 MAP kinases; b) c-Jun-N-terminal or JNK or stress-activated protein kinases (SAPK); c) p38 group: p38α, p38β, p38γ (ERK-6, SAPK3) and p38δ (SAPK-4); d) ERK-5, also called Big MAP kinase (BMK), and e) ERK-3 [8-14]. The first three mammalian MAPKs, ERK 1, ERK2, and ERK 3, were cloned in the early 1990s, then a second group of MAPKs, the JNK/SAPKs, was discovered as a cycloheximide-activated proline-directed kinase and as an activity that bound to and phosphorylated the N-terminal sites of c-Jun following exposure of cells to UV light [62,63]. The p38 subgroup was discovered as a lipopolysaccharide (LPS)-induced tyrosine phosphoprotein and as the target of a drug developed to inhibit LPS-induced tumor necrosis factor-biosynthesis [64,65]. ERK-5 was originally cloned as a homologue of ERK 1/2 and as a protein that interacted with the orphan MEK, MEK5 [66,67].

Unlike the ERK 1/2, which is activated in response to growth and differentiation factors, the JNKs and p38 MAPK are usually activated in response to inflammatory cytokines, such as tumor-necrosis factor (TNF) and interleukin-1 (IL-1) and a diverse array of cellular stresses including UV light, X-rays, hydrogen peroxide (H_2O_2), heat and osmotic shock, and withdrawal of growth factors [8,16,18-21]. In vascular smooth muscle cells (VSMCs), H_2O_2 activates p38 MAPK, Big MAPK, and JNK, but its effects in ERK 1/2 have been controversial, with some reports showing inhibition and others demonstrating stimulation [68-71]. The p38 MAPK and JNK activation by angiotensin II is inhibited by antioxidants (DPI, NAC), p22phox antisense, or overexpression of catalase [68,72]; while arachidonic acid-induced stimulation of JNK occurs via Rac-dependent H_2O_2 production [73]. Because arachidonic acid is produced in response to many vasoactive hormones, this may represent a common mechanism of activation. Moreover, although PDGF-induced ERK 1/2 phosphorylation is inhibited by incubation with catalases, angiotensin II activation of these enzymes is not [68,71,72,74].

Angiotensin II differentially activates the three major members of the MAPKs family, ERK 1/2, JNKs and p38 MAPK [75-77]. These kinases mediate angiotensin II-indirect vascular smooth muscle cell (VSMC) hypertrophy [78,79]. In VSMC, it has been shown that angiotensin II activates ERK 1/2 via redox-insensitive mechanism [71], whereas p38 MAPK is preferentially activated by angiotensin II through intracellular generation of reactive oxygen species (ROS), like H_20_2 [78]. Other studies have reported that MAPKs are significantly activated in hypertrophied heart, ballon-injured artery, and hypertensive vascular or renal diseases [80-86].

The ERK pathway is often involved in the regulation of cell growth and differentiation and involves the activation of Ras. ERK activation is an important event of T-cell activation, and is required for Th2 differentiation. Deficient Ras and ERK activation was reported to exist in clones that are anergized. ERK-1-deficient mice exhibited defective thymocyte maturation. On the other hand, mice deficient for this kinase were found deficient in TNF-α production when exposed to LPS. The macrophages from these mice exhibited selective ERK activation deficits [87].

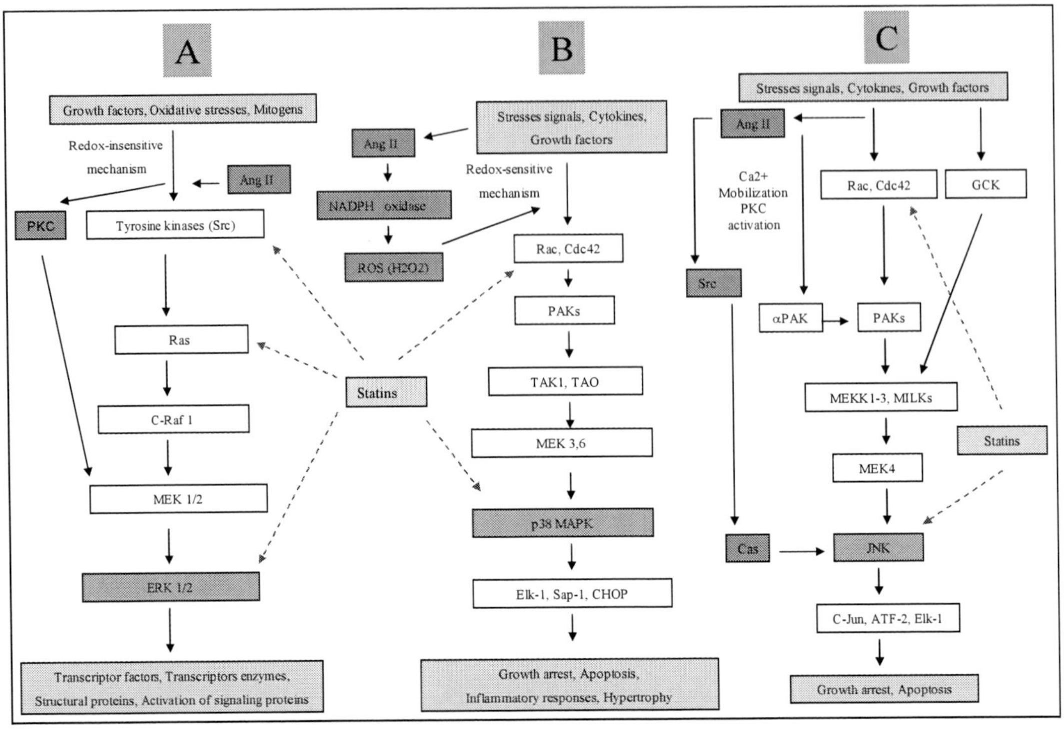

Figure 4. The MAP kinases cascades. A summary of three components of MAP kinases cascades, and the different known sites of statins actions are shown: A.ERK1/2 cascade; B. p38 MAPK cascade; C. JNK cascade.

The activation of p38 has been shown to be implicated in cellular responses including inflammation, cell cycle, cell death, development, cell differentiation, senescence, and tumorigenesis. The p38 pathway plays an essential role in the production of proinflammatory cytokines including IL-1β, TNF-α, IL-6, induction of COX-2, expression of iNOS, induction of VCAM and other adherent proteins, and regulation of the proliferation and differentiation of immune cells such as GMCSF, EPO, CSF, and CD40 [88-91]. Besides different studies have shown that p38 is necessary for IFN-γ expression in T-cells [87]. p38 MAPK also regulates activation-induced cell death of CD8+ cells, but not in CD4+ cells. Induction of death appears to be due to a reduction of Bcl2 levels selectively in CD8+ T-cells. Phosphorylation of c-Jun results in mitosis, differentiation and functions of peripheral immune cells such as T-lymphocytes or macrophages [92-95]. Observations from JNK knock-out mice demonstrated defined non-redundant functions of JNKs in the immune system. For example, JNKs trigger the depletion of positive T cells in the thymus, and most likely, this process is mediated by JNK2 which is responsible for the differentiation of progenitor cells into Th1-helper cells [92,93]. A central function of JNKs is the activation of the IL-2 promoter through AP-1 transcription and the stabilization of the IL-2 mRNA [94]. JNK2-deficient mice show dysfunctions of peripheral T-cells, and T lymphocytes without JNK1 or JNK2 reveal disturbances in the synthesis of cytokines [95]. JNK was described as a mediator of TRAIL, but not Fas-triggered apoptosis in T- and B-lymphocytes and the MHC-1 triggered apoptosis [96]. On the other hand, JNK2- deficient T-lymphocytes are not altered in the apoptotic response. Besides T-cells, JNK might be relevant for the differentiation of B-progenitor cells by IL-3, and for the IL-6 or TNF release from macrophages and mast cells [97,98]. In addition, functions of JNKs comprise the adhesion and infiltration of leucocytes mediated by expression of E-selectine in endothelial cells [99].

Specifically in SLE patients, it has been shown abnormal patterns of protein tyrosine phosphorylation and expression of key scaffold molecules such as the TCR ξ chain in human lupus T lymphocytes [100-105]. Cedeño et al (106), examined the activity of ERK1/2 and TCR-activated peripheral blood T lymphocytes from SLE patients, and also the binding of Ras guanine nucleotide exchange factor, human Son of Sevenless (hSos), to cytosolic adapter protein growth factor receptor-bound protein 2. They found that cells from SLE patients showed diminished catalytic activity and TCR-driven dual phosphorylation of ERK 1/2 upon stimulation through the TCR/CD3 receptor, a defect that may be related to altered translocation of hSos to the Ras/Raf membrane complex and diminished nuclear translocation of trans-acting factor AP-1

In accordance with these results, Deng et al [107] recently showed decreased activation of the Ras/MAPK pathway in human lupus T cells, a defect leading to deficient DNA methylation in this cell subpopulation. Conversely, Yi et al [108], found an increased expression of phosphorylated MAPKs associated with hyperexpression of CD40 ligand in human lupus T cells.

On the other hand, Rapoport et al (109), studied the expression, regulation and function of the p21Ras pathway in lymphocytes of 23 ambulatory SLE patients with active and non-active disease and eleven controls. Levels of p21Ras stimulatory element hSOS1 but not p21Ras and its inhibitory element p120GAP were significantly decreased in SLE patients. Early p21Ras signaling was down-regulated in SLE patients with active disease. Anchorage of p21Ras to the cellular membrane was also significantly decreased in these patients. In contrast, the late p21Ras signaling was up-regulated in SLE patients as indicated by the significantly higher constitutive activity of the p21Ras down stream key regulator enzyme MAPKs.

Collectively, the studies reviewed yield some molecular evidence on which to build the hypothesis that MAPKs may be involved in either cardiovascular and renal diseases, or autoimmune diseases and therefore statins through their effects on that pathways might be useful in the treatment of SLE and other rheumatic diseases.

STATINS, CARDIOVASCULAR SYSTEM AND MAPKS

Coronary heart disease (CHD), the result of coronary atherosclerosis, is the largest single killer of Americans and remains the leading cause of morbidity and mortality worldwide [110]. CHD was responsible for one in five or approximately 500.000 deaths in 1998 [111]. Elevated serum cholesterol was confirmed as a risk factor for CDH by Framingham study in 1984 [112]. Central to the pathogenesis of atherosclerosis are the deposition and retention of cholesterol in arterial walls, making lipid modification, the key to CDH prevention.

The HMG-CoA reductase inhibitors or statins, are the most commonly prescribed agents for the treatment of hypercholesterolemia, due to their efficacy in lowering LDL and ability to reduce clinical outcome in both primary and secondary prevention of coronary artery disease [45-49,113,114]. Recent evidence, however, suggests that the beneficial effects of statins may extend beyond their effects on serum cholesterol levels [115,116] possibly involving direct effects on the vascular wall [2,3,117-129]. These actions can be elicited by direct application of statins to cells in culture, as well as in animals and humans with normal cholesterol levels. These effects of statins seem independent of blood cholesterol lowering, and are known as the pleotropic actions of statins. Many of the pleotropic actions of the statins are in part due to their capacity to interfere with the mevalonate pathway on VSMCs and inhibit prenylation of Rho family GTPases and their translocation to the membrane with secondary inhibition of MAPKs pathways (Figure 3) [112,113].

On the other hand, the statins through their effects on small GTPase on vascular endothelial cells has been showed to attenuate endothelial MHC class II expression, increase endothelial nitric oxide synthase and fibrinolytic activity, decrease leukocyte adhesion and transmigration, and enhance resistance to local injurious stimuli [for review, see ref. 130]. All these effects could also contribute to prevent CVD in SLE patients.

This section will describe the effects of statins on MAPKs pathways on VSMCs and endothelial cells, and some of the immunomodulatory effects of statins, probably nonlipid mechanisms of cardiovascular protection, and provide an update review of new investigations that expand our knowledge on the pleotropic actions of statins that could be the base for understanding the beneficial cardiovascular effects of these drugs in SLE patients.

Statins and ERK Cascade

Activation of ERKs requires dual phosphorylation on threonine and tyrosine residues within the motif Thr-Glu-Tyr, which is mediated by MEK [131] (Figure 4A). This pathway can be activated by stimulation of receptors with intrinsic tyrosine kinase activity or G-protein-coupled receptors [132]. A number of agents with effects on smooth muscle cell hypertrophy and proliferation, vascular resistance, and platelet aggregation activate these kinases, include among these, the heparin-binding epidermal growth factor (HB-EGF), platelet-derived growth factor (PDFG), angiotensin II, endothelin (ET), thromboxane A_2, prostaglandin H_2, prostaglandin F_2, thrombin, norepinephrine, and acetylcholine [132-141]. Those agents can activate the ERK cascade though of biding different kind of receptors, like the epidermal growth factor receptor (EGF-R) and the PDFG receptor (PDFG-R), that serve not only as receptors for EGF and PDGF, respectively, but also as a scaffold for assembly of signalling complexes by G protein-coupled receptors such as those for angiotensin II [142,143]. Angiotensin II influences the activity of receptor tyrosine kinases, such as EGF-R, PDGF-R and IGF-R, even though it does not directly bind to these receptor tyrosine kinases. Mechanisms underlying angiotensin II-induced transactivation of receptor tyrosine kinases include activation of tyrosine kinases (PyK2 and Src) and redox-sensitive processes [144]. In addition angiotensin II stimulates phosphorylation of many non-receptor tyrosine kinases including PLC-γ, JAK (JAK and TYK), FAK, p130Cas and phosphatidylinositol 3-kinase (PI3K). EGF-R transactivation seems to be a Ca^{2+}-dependent process, whereas PDGF-R transactivation is Ca^{2+} independent. In response to G protein-coupled receptor agonists (ET-1, thrombin, carbachol, lysophosphatidic acid and tetradecanoyl-phorbol-13-acetate), HB-

EGF is generated by cleavage of pro-heparin-binding-EGF by metalloproteinase [145]. Free HB-EGF then binds to EGF-R resulting in EGF-R homodimerization and autophosphorylation. Similar processes have been demonstrated for IGF-R transactivation.

On the other hand, vascular endothelial growth factor (VEGF) is an endothelial cell-specific mitogenic and chemotactic agent. It induces in endothelial cells the production of the plasminogen activators required for the proteolytic degradation of the extracellular matrix during sprouting of capillaries and it increases the vascular hyperpermeability that precedes new blood vessel formation [146].

VEGF exerts its effects after binding with two receptor-tyrosine-kinases, Flt-1 (VEGFR-1) and KDR/Flk-1 (VEGFR-2), present at the surface of endothelial cells [147,148].VEGF receptors utilize signal transduction pathways that involve the activation of Ras GTPase activating protein, phosphatidylinositol-3 kinase (PI3-K), and phospholipase Cg (PLCg) [149-152].VEGF has been reported to activate ERK in capillary endothelial cells, rat liver sinusoidal endothelial cells and primary cultures of human umbilical vein endothelial cells, and this activation was associated with increased cell proliferation [150,153,154]. Activation of MEK1-ERK1/2 signaling is essential for VEGF-mediated proliferation and migration of endothelial cells [155].

Table 1. Comparation of the effects of different statins on basal phosphorylation of MAPKs on VSMCs

References	Statin	VSMCs	Concentration	Incubation Time	Effect
158	lovastatin	Bovine coronary	5 μmol/L	24 hours	Inh. ERK
159	simvastatin	A10 cells aortic, rat	10 μM	10-180 minutes	Inc. ERK, NC p38MAPK and JNK
162	cerivastatin	Human, aortic	0.1 μM	20 hours	Inh. ERK, NC JNK
166	fluvastatin	Sprague-Dawley rats	3 μM	24 hours	NC ERK and p38 MAPK
169	simvastatin	Human saphenous vein	0.5-10 μM	5 hours	NC ERK and p38 MAPK
170	cerivastatin	Human aorta	NR	24 hours	NC ERK and p38 MAPK
	lovastatin			24 hours	NC ERK
171	simvastatin	Sprague-Dawley rats	0.1, 1, 3 μM	24 hours	NC ERK and p38 MAPK
	lovastatin		0.5, 5 μM		

VSMCs: vascular smooth muscle cells; NC: not changed; NR: not reported; Inh: inhibition; Inc: increase

The different effects of statins on either basal or stimulated ERK cascade have been extensively studied with contradictory results (Table 1, 2) [156-171]. The inhibition on basal ERK for statins has been reported [158,162]. Lovastatin at a concentration of 5 μmol/L (24 hours) had a highly specific effect on ERK1/2 on bovine coronary smooth muscle cells. The basal phosphorylation of ERK1/2 was clearly reduced, but could be restored by addition of mevalonate, while the phosphorylation of ERK 1 was mildly suppressed and the phosphorylation of ERK 2 was reduced to non-detectable levels [163]. On the other hand,

Table 2. Comparation of the effects of different statins on stimulated phospholylation of MAPKs on VSMCs

References	VSMCs	Agonists	Concentratio/ Time	Statin	Concentration/ Time	Effect	Reversal
156	A10 cells (Aortic, rat)	PDFG-BB	30 ng/mL / 10 minutes	pravastatin	100 μM / 1hour	NC ERK	N/A
157	Human mammary artery	PDFG	2,5,10,20 ng/mL / 10 minutes	simvastatin	5 μM / 25 hours 10 μM / 2 and 24 hours	NC ERK NC p38 MAPK	N/A
158	Bovine coronary	bFGF	10 ng/mL / 10,30,90 min	lovastatin	5 μmol/L/ 24 hours	Inh. ERK	ERK 1 Mev
160	Sprague-Dawley rats, aorta	LysoPC	12.5 μM / 7 minutes	pitavastatin	10-100 μM*/ 48 hours	Inh. ERK	Mev, FPP
				atorvastatin	24 hours	Inh. ERK	NR
				fluvastatin	24 hours	Inh. ERK	NR
161	Sprague-Dawley rats	Ang II	10^{-7} mol/L / 1 minute	cerivastatin	5 μmol/L / 20 minutes and 6,12,24 hours	Inh. ERK	NR
163	Male C57BL/6J mice	Ang II	1 μmol/L / 15 minutes	fluvastatin	0.1-5.0 μmol/L / 24 hours	Inh. ERK	NR
164	Male Wistar rats	FBS	10% / 5 days	fluvastatin	1 μmol/L / 5 days	Inh. ERK and p 38 MAPK	Mev
165	Human aorta	HDL	30 mg/dL / 5 minutes	simvastatin	0.1 and 1μM / 24 hours	NC ERK 2, JNK and p 38 MAPK	N/A
166	Sprague-Dawley rats	IL-1 beta	5 ng/mL / 10 minutes	fluvastatin	3 μM / 24 hours	NC ERK and p 38 MAPK	N/A
167	Bovine coronary	bFGF	2 ng/mL / 15 minutes	lovastatin	10 μmol/L / 18 hours	Inh. ERK	NR
168	A7r5 (embryo rat thoracic aorta)	PDGF-BB	20 ng/mL / 10 minutes	cerivastatin	2 μM / 6 hours	Inh. ERK	GGPP
169	Human Saphenous vein	PDGF + IL-1	15 ng/mL + 20 ng/mL/ 5 minutes	simvastatin	0.5-10 μM / 5 hours	NC ERK and p38 MAPK	N/A
170	Human aorta	Thrombin	1 U/mL / 5 minutes	cerivastatin	100 nmol/L / 24 hours	NC ERK, Inh. p 38 MAPK	NR
171	Sprague-Dawley rats	Ang II	1 nM/ 10 minutes	simvastatin lovastatin	3 μM 5 μM	Inh. p38 and ERK NC ERK and p38	NR
174	Human aorta	D-glucose	25 mmol/L / 24 hours	lovastatin mevinolin	NR 10 μmol/L/24 hours	NC ERK Inh. ERK	NR
189	Human aorta	H_2O_2	4 μM / 1 hour	simvastatin	10 μM / 24 hours	Inc. p38 MAPK	Mev

VSMCs: vascular smooth muscle cell; NC: not changed; NR: not reported; Inh: inhibition; Inc: increase; *100 for Pitavastatin, 10-50 for Fluvastatin, 10-100 for Atorvastatin; GGPP: geranylgeranylpyrophosphate; FPP: farnesylpyrophosphate; Ang: angiotensin; Mev. Mevalonate.

cerivastatin at a concentration of 0.1 μM (20 hours) inhibited basal ERK 1/2 phosphorylation in human VSMCs [162]. Also, the enzymatic activity of ERK was blocked by either PD98059 (a specific inhibitor of MEK1/MEK2) or cerivastatin in a dose-dependent manner, with additive effects by the combination [162]. Other authors have shown not effect on basal ERK phosphorylation by statins [166,169-171]. Cerivastatin (24 hours incubation) per se had only a small effect on the low basal ERK 1/2 phosphorylation and no effect on the expression of this kinase [171] while that fluvastatin at a concentration of 3 μM showed no effect on basal activity of ERK1/2 [166]. Similar results were obtained with simvastatin and lovastatin [169,171]. Conversely, simvastatin at a concentration of 10 μM induced a significant phosphorylation of ERK1/2 [159]. Simvastatin-induced phosphorylation of ERK1/2 reached its peak at 60 min after the stimulation. This effect of the simvastatin at a concentration of 20 μM (3 hours) was markedly reduced either by pre-treatment for 60 minutes with PD98059 (10 μM) or U-0126 (10 μM), both of them specific inhibitors of upstream kinases [172,173].

The inhibition of stimulated ERK phosphorilation has been more extensively reported [158,160,161,163,164,167,168,171,174]. Lovastatin at a concentration of 5 μmol/L (24 hours), clearly reduced the stimulated (10 ng/mL of basic fibroblast growth factor (bFGF); 10, 30 and 90 minutes) phosphorylation of ERK1/2 [158]. While the phosphorylation of ERK 1 could partially be restored by addition of mevalonate, the reduced phosphorylation of ERK 2 could not be restored by addition of excessive doses of it or stimulation of the cells with bFGF [158]. Conversely, lovastatin at a concentration of 10 μmol/L (18 hours) only partially inhibited bFGF-induced phosphorylation of ERK1/2 after 15 minutes of stimulation [167]. In contrast, MAPK phosphorylation was sustained after 5 hours of stimulation in lovastatin-pretreated cells. They previously observed that bFGF (2 ng/ml) caused a rapid phosphorylation of ERK1/2 after 15 minutes, and could be observed after 1 hour, but was back to basal levels after 5 hours. In other study, Yamakawa et al [160], found markedly inhibited lysoPC-induced ERK 1/2 phosphorylation in VSMCs incubated with pitavastatin, atorvastatin or fluvastatin at a concentration of 100 μM (24-48 hours). The inhibition effect of these statins was concentration-dependent. The co-incubation of VSMCs with mevalonate partially stopped the inhibition of ERK 1/2 phosphorylation by pitavastatin. Similarly, co-incubation with farnesyl pyrophosphate (FPP) partly prevented the effect of pitavastatin on ERK 1/2 phosphorylation. Cerivastatin and fluvastatin, others statin that has been tested, had showed to inhibit the angiotensin II-induced phosphorylation of ERK 1/2 [161,163]. Cerivastatin at a concentration of 5 μmol/L (20 minutes, 6, 12 and 24 hours) inhibited angiotensin II-induced phosphorylation of ERK1/2 [161]. While, fluvastatin at a concentration of 0.1 to 5 μmol/L (24 hours) significantly inhibited the ERK 1/2 activation induced by angiotensin II (1 μmol/L for 15 minutes) [163]. Consistent with these in vitro results, these authors observed that activation of ERK 1/2 was enhanced in response to vascular injury 7 days after surgery. Administration of both fluvastatin and valsartan decreased the activation of ERK 1/2, without change of total protein level of ERK 1/2 induced by vascular injury, whereas fluvastatin and valsartan alone at these doses did not affect these parameters. Similarly, fluvastatin at a concentration of 1 μmol/L (5 days of incubation) produced a large inhibitor at the FBS-induced ERK 1/2 phosphorylation (previously, the authors demonstrated that 10% FBS significantly increase ERK 1/2 phosphorylation) [164]. In addition, 100 μmol/L mevalonate treatments completely prevented the effects of fluvastatin. Farnesyl protein transferase inhibitor III (FPT III)(25 μmol/L) completely inhibited the FSB-induced ERK 1/2 phosphorylation, but mevalonate treatment did not restore the inhibitory action of FPT III. Also, geranylgeralnyl transferase inhibitor (20 μmol/L) had no effect on FBS-induced ERK 1/2 phsophorylation. Kamiyama et al [168], showed that cerivastatin at a concentration of 2 μM (6 hours) inhibited the PDGF-BB-activation of ERK 1/2 in VSMCs. This effect was completely reversed by the addition of geranylgeranyl pyrophosphate (GGPP), but the addition of farnesyl pyrophosphate (FPP) had no effect. Finally, has been shown that mevinolin at a concentration of 10 μmol/L (24 hours) inhibited the glucose-induced ERK1/2 phosphorylation and activation [174]. A significant increase in ERK 1/2 phosphorylation was seen with 25 mmol/L glucose (24 hours). Recently, has been showed that simvastatin (0.3, 1, and 3 μM) inhibited angiotensin II-mediated phohorylation of ERK 1/2 [171].

Only few reporters have shown no effects of statins on stimulated ERK 1/2 phosphorilation [156,157,165,166,169-171]. Pravastatin at a concentration of 100 μM (1 h) showed no diminution on PDFG-BB-stimulated ERK1/2 activity (30 ng/ml for 10 minutes) on VSMCs [156]. However, the MAP kinase activation was strongly induced by PDGF-BB. Probably their results were due of short-term pravastatin incubation. Similar results were observed with simvastatin at a concentration of 5 μM (24 hours) [157]. Simvastatin failed to inhibit ERK 1/2 phosphorilation at all PDFG concentrations tested (1,5,10,20 ng/ml) [157]. HDL (30 mg/dl) and thrombin (1 U/mL) induced a marked increase in ERK 1/2 phosphorylation [165,170]. Those effect were not affected by cerivastatin or lovastatin treatment (24 hours incubation) [170] or simvastatin-treated (0.1 and 1 μM) cells on ERK 2 phosphorylation [165].Similarly, fluvastatin at a concentration of 3 μM showed no effect on IL-1 β-stimulated activity of ERK 1/2 [166]. The induced activation of ERK 1/2 by the combination of 15 ng/ml PDGF plus 20 ng/ml IL-1α (PDGF/IL-1) (5 minutes) was also not affected by simvastatin at a concentration of 0.5 to 10 μM (5 hours) [169]. Finally, lovastatin at a concentration of 0.5 and 5 μM did not significantly reduced phosphorylation of ERK 1/2 stimulated by angiotensin II [171].

Statins and p38 MAPK Cascade

The p38 MAPK activation has been observed in response to a variety of extracellular stimuli in different organisms [19] (Figure 4B). Mammalian p38 MAPK activation has been shown to occur in response to extracellular stimuli such as UV light, heat, osmotic shock, inflammatory cytokines (TNF, IL-1), growth factors (CSF-1) [175-184], and by agonist that play important roles in hypertension including angiotensin II, ET-1, and α–adrenergic agents [68,72]. Like ERK, the p38 MAPK is proline directed and require phosphorylation on both tyrosine and threonine residues for activation [184]. p38 MAPK has a TGY motif within kinase subdomain VIII which, when is phosphorylated, activates the kinases. There is 40% to

50% identify in catalytic domains when comparing the ERK and p38 MAPK [183]. Angiotensin II phosphorylates vascular p38 MAPK, which plays an important role in inflammatory responses, apoptosis and inhibition of cell growth [20,185,186]. In the cardiovascular system, p38 MAPK has been implicated in cardiac ischemia, ischemia/reperfusion injury, and arterial remodeling in hypertension [20]. The specific cascade regulators of angiotensin II-activated p38 MAPK in VSMCs are unclear, but p38 MAPK could be a negative regulator of ERK 1/2 [185].

Recently, has been showed that VEGF, increases COX-2 gene regulation at both mRNA and protein levels in cultured human vascular endothelial cells through p38 MAPK [187]. The activation of p38 MAPK in primary cultures of human umbilical vein endothelial cells (HUVECs) also mediates actin organization and cell migration as an important modulator of the angiogenic effect of it [154].

The phosphorylation of tyrosine 1214 on VEGFR2 is required to trigger the sequential activation of Cdc42 and SAPK2/p38 and to drive the SAPK2/p38-mediated actin remodeling in stress fibers in endothelial cells exposed to VEGF [188].

The effects of statins on p38 MAPK has been less extensively studied than on ERK 1/2, but with the same contradictory results (Table 1,2). The statins have been shown to inhibit [164,170,171], increases [189], or no affect [157,159,165,166,169-171] either the basal or the stimulated p38 MAPK phosphorilation. Cerivastatin at a concentration of 100 nmol/L (24 hours) and fluvatatin at a concentration of 1 μmol/L (5 days) inhibited the thrombin and the FBS stimulated phosphorylation of p38 MAPK respectively [164,170]. Thrombin (1 U/mL, 5 minutes) and FBS induced a marked increase in p38 MAPK phosphorylation [164,170]. However, the H_2O_2 (4mM) induced p38 MAPK phosphorylation was enhanced by pre-incubation with 10 μM simvastatin for 24 hours [189]. Co-addition of mevalonate prevented this phosphorylation, indicating that isoprenylated proteins are probably involved in this process. The same result was obtained with SB203580, a specific inhibitor of p38 MAPK. Finally, simvastatin (0.3, 1, and 3 μM for 24 hour) inhibited angiotensin II-mediated phohorylation of p38 MAPK [171].

Most of the papers published show not effect of the statins on p38 MAPK phosphorilation [157,159,165,166,169-171]. Simvastatin at a concentration of 10 μM (10,20,30,60,90,120, and 180 minutes) did not affect the phosphorylation of basal p38 MAPK [159]. Similar results were observed with cerivastatin at a concentration of 100 nmol/L (24 hours) [170], fluvastatin at a concentration of 3 μM [166] (24 hours), simvastatin at a concentration of 0.5 to 10 μM (5 hours, and 24 hours) [169,171], and lovastatin at a concentration of 0.5 and 5 μM [171]. On the other hand, simvastatin at a concentration of 10 μM (at variables time) failed to inhibit PDFG-induced p38 MAPK phosphorilation after 2 and 24 hours of pretreatment [157]. The p38 MAPK phosphorylation was strongly induced by PDFG at concentrations as low as 2 ng/ml. Similar results were observed with simvastatin at a concentration of 0.1 or 1 μM on HDL-induced p38 MAPK phosphorylation [165], and simvastatin at a concentration of 0.5 to 10 μM (5 hours) on the induced activation of p38 MAPK by the combination of 15 ng/ml PDGF plus 20 ng/ml IL-1α (PDGF/IL-1) (5 minutes) [169]. Finally, fluvastatin at a concentration of 3 μM [166] (24 hours) showed no effect on IL-1β stimulated activity of p38-MAPK [166].

Statins and JNK Cascade

Like ERKs and p38 MAPK, the JNK is proline directed and requires phosphorylation on both tyrosine and threonine residues for activation [184], but JNK contain a TPY motif (Figure 4C) [190]. JNK is activated by the same cellular stresses that p38 MAPK. However, differences in activation patterns do exist. For example, JNK cascade is not activated by ischemia alone, but is markedly activated by reperfusion of ischemic kidney [191,192]. JNK is also activated in the arterial wall of aortic, carotic, and femoral arteries by acute hypertension in rats whether the hypertension is caused by angiotensin II or phenylephrine infusion [193]. Angiotensin II in liver epithelial cells, and endothelin, in airway smooth muscle cells and glomerular mesangial cells, activated JNK and, of note, JNK activation is greater than ERK activation [194-196]. These data raise the possibility that JNK may be a major signaling arm of the vasoactive peptides and suggest that these kinases might play a role in the hypertrophic adaptation of myocytes, VSMCs, and renal mesangial cells [197].

Angiotensin II activates JNK, which regulate VSMC growth by promoting apoptosis or by inhibiting growth [21,76,198,199]. Angiotensin II phosphorylates JNK via p21-activated kinase (αPAK), which is dependent on intracellular Ca2+ mobilization and on PKC activation [199]. Following phosphorylation, the isoforms JNK-1 and JNK-2 traslocate to the nucleus to activate a number of transcription factors, such as c-Jun, ATF-2, and Elk-1 [21]. Recently have been shown that angiotensin II also stimulate JNK via Src-Cas pathway [200].

The effects of statins on JNK cascade have been poor studied (Table 1,2) [159,162,165]. Simvastatin at a concentration of 10 μM (10,20,30,60,90,120, and 180 minutes), and cerivastatin at a concentration of up to 10 μM (20 hours) did not affect the phosphorylation of basal JNK [159,162]. Beside, the HDL (30 mg/ dl) induced phosphorylation of JNK was no affected by treatment with simvastatin at a concentration of 0.1 or 1 μM on VSMCs [165].

In summary, the different effects of the statins on the three major MAPKs cascades (ERK1/2, p38 MAPK, and JNK) on VSMCs have been studied, but with contradictory results. However, there are differences among the methods used in those studies (Table 1,2). I found that simvastatin increased [189], did not change[157] or inhibited [171] the stimulated phosphorylation of p38 MAPK. In two out of three cases were used human VSMCs, but in al cases were used the same incubation time with simvastatin, however, were used different agonists (PDFG, H_2O_2, angiotensin II) and simvastatin concentration (5 and 10 μM). All the agonist used increased the phosphorylation of p38 MAPK, but how the simvastatin enhanced the oxidative stress-induced p38 MAPK phosphorylation, remain to be answered. The only consistent result with simvastatin was its effect on stimulated phosphorylation of ERK 1/2 in two occasions, which was not changed [157,165]. In those cases, also were used different agonists (PDFG and HDL), and low doses of simvastatin (≤ 5 μM), but the conditions were similar (24 hours of incubation, and same VSMCs). However, in a personal experience [171] I found that simvastatin (3 μM during 24 hours) inhibited the angiotensin II-induced ERK phosphorylation on VSMCs from rats. Probably, the simvastatin has a stimulation effects on p38 MAPK phosphorylation, when is used in higher doses. Fluvastatin was other statin with differents effects on MAPKs cascades [160,163,164,166]. In three of those studies [160,163,164], fluvastatin inhibited the stimulated phosphorylation of ERK 1/2, while that the other study reported no effect on this kinase [166]. All of those studies used different agonists, doses of the fluvastatin, incubation times, and cells types. However, the results were contradictory even with comparable conditions, as same VSMCs and similar time of fluvastatin incubation. This underlines the potential role the different statins-sensitive, prenylated proteins, that could be activates for different agonists, in the regulation of MAPKs pathways.

The same contradictory results were obtained with fluvastatin and p38 MAPK. Fluvastatin in lows doses (≤ 3μM), either inhibited or did not change the stimulated p38 phosphorylation [164,166]. In this case, 5 days of fluvastatin incubation inhibited stimulated p38 phosphorylation [164], while that no change in the p38 phosphorylation was observed after 24 hours of fluvastatin incubation [166]. Probably, this effect was due to a longer incubation time. These contradictory results also have been seen with cerivastatin. In similar conditions, but in different VSMCs, this statin inhibited the angiotensin II-induced ERK1/2 phosphorylation [161], but did not change the thrombin-induced ERK 1/2 phosphorylation [170]. The effects of the statins on basal MAPKs phosphorylation have been poorly studied. However, the results are also contradictory [162,171].

Beside of the differences within of the same statin, there are important differences among statins class (Table 1,2). This, in part, could be due to their solubility differences (more or less hydrophilic or lipophilic).

Statins could be acting in different levels of the MAPKs cascade (Figure 2). I found that the statins affect the tyrosine kinases phosphorylation, Rac, Ras, and Raf expression, and MAPKs phosphorylation. However, the exact mechanism of their effects on the MAPKs cascade is not known, so far.

Immunomodulatory Effects of Statins and Atherosclerosis

Although, central in the pathogenesis of atherosclerosis are the deposition and retention of cholesterol in arterial walls, immunological mechanisms play an important role in both the onset and the development of atherosclerosis, in converting a stable atherosclerosis plaque into an unstable one and in overlapping thrombosis that occurs [201-206]. Different studies have shown the presence of macrophages, monocytes and T lymphocytes in the atheroma plaques, as well as on the observation of the presence of C-reactive protein (CRP), components of the complement, and serum A amyloid protein inside these plaques [201,207].

Aprahamian et al [208], developed a murine model crossing a mouse strain susceptible to atherosclerosis (apoE -/-), with a strain that develops lymphoproliferation and autoimmunity (gld), characterized by an impaired Fas-mediated apoptosis. In these gld.apoE -/- mice the atherosclerotic lesions were significantly increased compared with apoE -/- mice.

Systemic lupus erythematosus (SLE) is an inflammatory rheumatic disease of immunologic origin characterized by autoantibody production. SLE has now become a chronic disease with 5-year survival rates of 90% or better [37,209,210]. Patients who survive the early years are at risk for accelerated CVD [37-42].

This is particularly striking in young women; Manzi et al [42], estimated that the incidence of acute myocardial infarction in women with SLE aged 35–44 yr was 50-fold greater than that of the comparable cohort in the Framingham Offspring Study. Young women without SLE are usually free from atherosclerosis. In individuals with SLE the prevalence of CVD ranges from 6% to 10% [211] and the risk of developing CVD is 4–8 times higher than in the normal population. Moreover, acute MI is reported as a cause of death in 3–25% of individuals with SLE [212].

Esdaile et al [43] estimated a 17-fold risk of mortality secondary to coronary heart disease, a 10-fold risk of non-fatal myocardial infarction and an 8-fold increase of stroke.

Studies using carotid ultrasound and echocardiography, have reported an odds ratio for cardiovascular disease in SLE compared with controls of 4.8, and have also revealed carotid plaques in 32% of 214 women with lupus [213,214]. Svenungsson et al [205] performed ultrasonographic measurements of the common carotid artery in 26 individuals with SLE and pre-existing CVD (SLE cases), 26 individuals with SLE but without pre-existing CVD (SLE controls), and 26 population controls. They found plaques in 65% of SLE cases, 38% in SLE controls and in 11% of the population controls. Some factors like osteoporosis, lupus anticoagulant, higher cumulative dosage of steroids, high levels of triglycerides, alfa-1 antitripsin, oxLDL, anti-oxLDL, lipo protein A, homocysteine and low levels of high-density-lipoprotein cholesterol were found more often in SLE cases than SLE controls. This study found no relationship between plaques and SLE disease variables, including renal involvement.

Doria et al [216], studied 78 individuals with SLE but without overt CVD, and found that an interaction between age and cumulative prednisolone intake seemed to be relevant for atherosclerosis in SLE patients.

The pathogenesis of CVD in patients with SLE appears to be multifactorial [211,217-220]. Even when studies control for classical cardiac risk factors, including hyperlipidemia, hypertension, and sedentary lifestyle, SLE is associated with excess risk.

Those factors are all prevalent in patients with SLE, but can not explain by themselves the magnitude of this increased risk; therefore the duration and activity of SLE itself confer excess risk of CVD [43,213,221]. The CRP, a soluble inflammatory mediator, has been associated with cardiovascular risk in the general population [222]. The LUMINA study of cardiovascular disease in patients with lupus, CRP still emerged as a risk factor; the median CRP in all patients with vascular events was 12.6 mg/l compared with a median CRP of 4.8 mg/l in patients without vascular events [223].

On the other hand, the prevalence and/or severity of traditional risk factors are increased in SLE [216,223]. For example, dyslipidemia has been demonstrated in SLE, prevalence of insulin resistance is increased in these patients and a risk-associated body habitus, with abdominal obesity, and hypertension is clearly increased in SLE [224,225]. Adipose tissue itself produces proinflammatory cytokines, or adipokines, and a relationship between insulin resistance and subclinical inflammation has recently been proposed [226].

It is probable that the interaction between endothelial cell dysfunction due to soluble inflammatory mediators and antibodies with the traditional risk factors, contributes to the development of premature atherosclerosis and results in an increased prevalence of cardiovascular disease. The fact of that the immune system is involved in the pathogenesis of atherosclerosis provides the potential for novel methods to treat or prevent the development of this disease in SLE patients.

Statins reduce atherosclerosis by lowering cholesterol levels but may also benefit lupus disease itself. As result, amelioration of autoimmunity by statins will lead to reduced disease activity and as a direct consequence less atherosclerosis.

Results of epidemiologic studies of methotrexate therapy are commensurate with this hypothesis, as is evidence suggesting that suppression of inflammation improves endothelial-dependent function [227-229]. Recently, Sattar et al [230], demonstrated for first time that targeting the TNF pathway can significantly decrease Lp(a) and homocysteine levels and elevate Apo A-I and sex hormone binding globulin (SHBG) concentrations. These data support an important precursor role for high-grade inflammation in modulating these putative risk parameters.

Statins would interfere with T cell activation via inhibition of normal recruitment of TCR and co-stimulatory molecules by depleting membrane cholesterol and disrupting the ordered architecture of molecules forming the lipid rafts of the cell membrane (Figure 5) [236-238]. It has been demonstrated that mice with depletion of CD4 cells reduce the formation of atherosclerosis lesions, and others that present cellular immunodeficiency, when they receive CD4 cells transferred from immunocompetent animals, present increase of atherosclerosis [234,235].

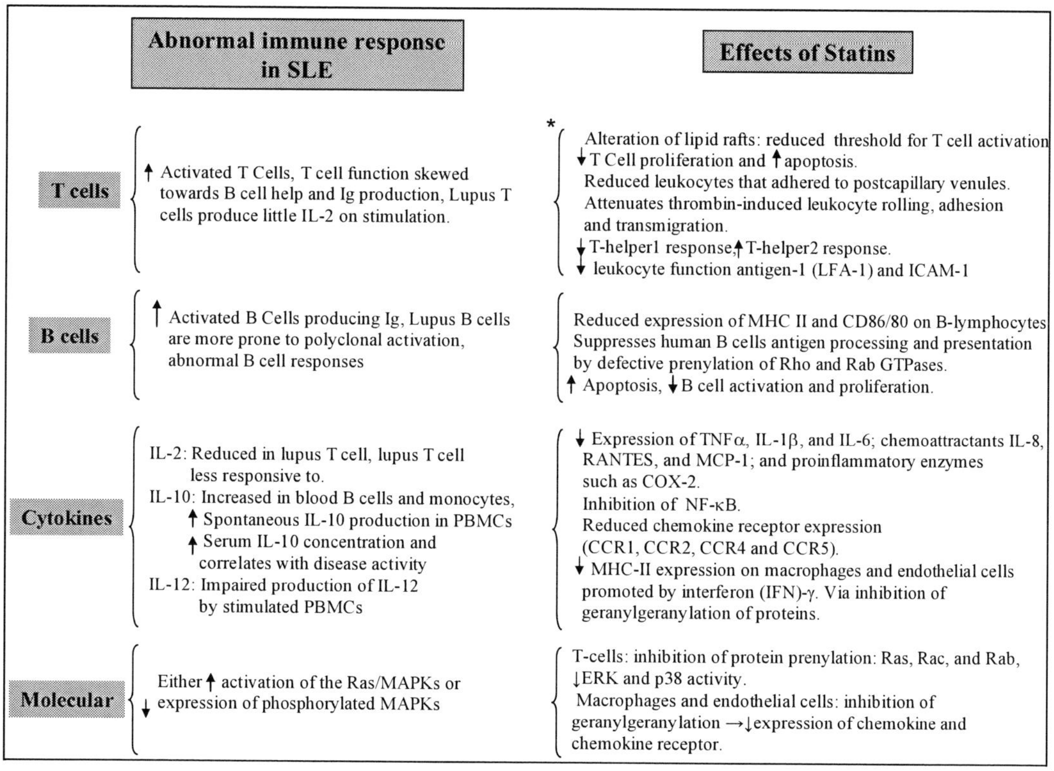

Figure 5. Summary of abnormal immune response in patients with SLE (for review, see ref. 270), and the potentially beneficial effects of statins in those abnormalities. * Effects elicited through the impairment in the pathways regulated by small GTPases, including the Ras/MAP kinase pathway, the Rac/stress kinase pathway.

Disruption of lipid raft domains by cholesterol depletion can disrupt immune cell signaling pathways. Cholesterol depletion using drugs such as methyl-β-cyclodextrin, has been shown to induce unregulated protein tyrosine phosphorylation in unstimulated cells and reduced Ca+2 mobilization and receptor aggregation in T cells activated using antibodies to CD3 [236]. Furthermore, cholesterol depletion using filipin and nystatin resulted in the inhibition of anti-CD3-stimulated phosphorylation of proximal signalling proteins [237].

Lovastatin has been shown to inhibit protein tyrosine phosphorylation in Jurkat T cells stimulated with antibodies to CD3 [238] and fluvastatin inhibits lipid raft dependent Fcg receptor signaling in monocytes [239].

The statins through this mechanism could target abnormalities in the expression and composition of lipid rafts that contribute to T cell dysfunction. These abnormalities include a reduced threshold for T cell activation, responsible for the heightened sensitivity and the prolonged response to activation [240].

On the other hand, endothelium perturbation has been demonstrated in systemic autoimmune diseases as supported by the increased plasma levels of molecules released or secreted by activated/damaged endothelial cells or by the impaired endothelium-dependent functional vessel wall responses [241,242]. It has been demonstrated that statins attenuate endothelial MHC class II expression, increase endothelial nitric oxide synthase and fibrinolytic activity, decrease leukocyte adhesion and transmigration, and enhance resistance to local injurious stimuli. Many of these effects are brought about by the modulation of small GTPase function and the downregulation of proinflammatory gene expression [for review, see ref. 130].

In summary, the pathogenesis of CVD in patients with SLE is multifactorial. There is a participation of classical cardiac risk factors, including hyperlipidemia, hypertension, and sedentary lifestyle, all of them increased in SLE patients, associated to immunological mechanisms which contributes to the development of premature atherosclerosis and results in an increased prevalence of cardiovascular disease. Therefore, the statins through both their serum lipid-lowering action, and their pleotropic actions including immunomodulatoy and anti-inflammatory effects, have the ability to reduce clinical outcome in both primary and secondary prevention of coronary artery disease. The pleiotropic effects of statins result, in part from their capacity to interfere with the mevalonate pathway and inhibit prenylation of Rho family GTPases and MAPKs phosphorylation.

STATINS, IMMUNOLOGIC SYSTEM AND MAPKS

The pleiotropic effect of statins includes significant anti-inflammatory and immunomodulatory actions [22-30]. In part, the anti-inflammatory effects of statins can result from their capacity to interfere with the mevalonate pathway and inhibit prenylation of Rho family GTPases and ultimately in MAPKs activation [31-36]. As was mentioned above, the MAPKs (ERK, p38, JNK) participate in the activation and differentiation of different immune cells (Figure 5). There are few studies where has been studied the effects of statins in SLE. Abud-Mendoza et al [243], showed that simvastatin (80 mg QD for 8 days) induced a rapid and significant reduction in proteinuria levels in three SLE patients. More recently, a study has shown that atorvastatin (30 mg/kg for 14 days) could delay the progression of established autoimmune disease in the NZB/W spontaneous murine model of SLE [244]. Treatment with atorvastatin resulted in a significant reduction in anti-dsDNA antibodies, proteinuria, glomerular Ig deposition and glomerular hypertrophy. These effects were along by a significantly reduced expression of major histocompatibility complex (MHC) II and CD86/80 on B-lymphocytes and consequently autoreactive T-cell proliferation was profoundly impaired.

Effects of Statins in other Rheumatic Diseases

There is a marked absence of clinical trials to confirm the anti-inflammatory effects of statins observed *in vitro* and in animal studies [243,245,246]. Abud-Mendoza et al [243], showed also that simvastatin (80 mg QD for 8 days) induced a marked beneficial effect in a patient with Wegener's granulomatosis and a patient with erythema nodosum. On the other hand, five RA patients who received atorvastatin (20mg QD for 8 days) showed reduction in C-reactive protein levels and a clinical improvement that was classified as an ACR20 response. Also, they showed in an open clinical trial of 15 patients with RA who received methotrexate as a single DMARD with no satisfactory response that simvastatin (40 mg QD) had an ACR50 or better response after 8 weeks in most patients (9/10). In comparison, no such response was observed in any patient (0/5) treated with chloroquine. McCarey et al [245], in a randomized double-blind placebo-controlled trial, found that in RA patients treated with atorvastatin (40 mg QD) during 6 months, the DAS28 improved significantly compared to placebo. DAS28 EULAR response was achieved in 18 of 58 (31%) patients compared with six of 58 (10%) allocated placebo. C-reactive protein and erythrocyte sedimentation rate (ESR) declined by 50% and 28%, respectively, relative to placebo. Swollen joint count also decreased. Kanda et al [246] in an open-label 12-week study treated RA patients with 10 mg per day of simvastatin and reported that simvastatin significantly reduced the number of tender joints, and patient self-assessment of disease activity declined on visual analog scale (VAS). Additionally, the ESR and rheumatoid factor (RF) were significantly reduced and the C-reactive protein showed a tendency to decrease. Finally, Ten Cate et al [247], successfully treated a therapy-refractory systemic juvenile idiopathic arthritis patient with 30 mg per day of atorvastatin along with steroids, which were tapered and discontinued without documented adverse events.

Animal and In Vitro Studies

Other evidences about the anti-inflammatory benefits of statins come from animal studies [248-251]. It has been shown that lovastatin (doses of 5-10 mg/kg) inhibit leukocyte recruitment in an animal model of acute inflammation. This inhibition was probably associated with the down-regulation of RANTES (regulated upon activation normal T-cell expressed and presumably secreted), monocyte chemotactic protein-1 (MCP-1) and interleukin 6 (IL-6) [248]. On the other hand, fluvastatin treatment (6 mg/kg) of hypercolesterolemic rats reduced the number of leukocytes that adhered to postcapillary venules in response to platelet-activating factor or leukotriene B4 [249] and rosuvastatin (0.5-1.25 mg/kg) attenuates thrombin-induced leukocyte rolling, adhesion and transmigration [250]. The direct anti-inflammatory effect of statins has been demonstrated using an established model of acute inflammation (carrageenan-induced foot pad edema) [251]. These authors demonstrated a significant reduction in edema formation following oral simvastatin treatment (3-10 mg/kg). Statins have also shown to have immunoregulatory effect in different autoimmune diseases, due to reduced MCH-II expression, and in a shift from the pathogenic T-helper1 (Th1) response to a protective T-helper2 (Th2) response [25, 252-254].

Studies in rodents have shown that *in vivo* lovastatin treatment reduced infiltration of mononuclear cells in the central nervous system parenchyma in experimental allergic encephalomyelitis [255].

The statins exert immunodolulatory effects through of their capacity to inhibit expression of proinflammatory proteins, including TNFα, IL-1β, and IL-6; chemoattractants IL-8, RANTES, and MCP-1; and proinflammatory enzymes such as COX-2 [22,29,256]. Further, statins have been implicated to inhibit of NF-κB and to increase the activity of transcriptions factors, which antagonize proinflammatory gene expression, including Oct-1 and the PPAR family transcription factors [22,29]. Other immunomodulatory effects of the statins involve suppression of T-cell activation and apoptosis [23-25,27,30,256]. It has been shown that statins can inhibit inducible MHC-II expression on macrophages and endothelial cells promoted by interferon (IFN)-γ [30], and directly bind to leukocyte function antigen-1 (LFA-1), a molecule that promotes leukocyte adhesion and delivers a strong co-stimulatory signal essential for T-cell activation [23]. Statin binding prevents the interaction of LFA-1 with its ligand ICAM-1 that locks the receptor in an inactive conformation and modulates T-cell differentiation, resulting in suppression of antigen specific T-cell activation [24,25]. Beyond the effects on LFA-1, statins may have a broader array of molecular targets in T-cell activation, including the TCR signal cascade through of its inhibitory effects on Ras and Ras-related GTPases [26,27]. Simvastatin has been recently shown to inhibit protein prenylation on T-cell, resulting in a dramatic impairment in the pathways regulated by small GTPases, including the Ras/MAP kinase pathway, the Rac/stress kinase pathway, and the Rab-dependent pathway of receptor endocytosis [27]. Simvastatin also reduced chemokine and chemokine receptor expression (CCR1, CCR2, CCR4 and CCR5), in human endothelial cells and macrophages via inhibition of geranylgeranylation of proteins [256].

Simvastatin potently suppresses tetanus toxoid processing and presentation by human B cells and dendritic cells to $CD4^+$ T cells by HLA-DR by inhibiting protein antigen uptake through both receptor-mediated endocytosis and macropinocytosis. This effect can be largely accounted for by defective prenylation of Rho and Rab GTPases in the absence of any measurable perturbation of lipid rafts. In addition, simvastatin was found to preferentially affect the invariant chain-dependent MHC class II pathway, thereby identifying this route of antigen processing and presentation as a selective target of statins [257]. Statins also are capable of eliciting significant inhibition of mitogen-stimulated human B-lymphocyte proliferation in a dose-dependent fashion, in both high-density, resting B lymphocytes and in B-leukemic cells which are susceptible to surface immunoglobulin triggering [258].

Apoptosis in several kinds of immune cell can be stimulated by statins [259-265]. Recently, Samson et al [259], reported that fluvastatin induced apoptosis in resting $CD4^+$ T-cells. Additionally, fluvastatin has been shown to cause an inhibition of normal T-cell proliferation in response to cross-linking of CD3 and reversed by mevalonate treatment. This effect was associated with reduction in ERK and p38 MAPK activity, consistent with inhibition of isoprenylation of G-protein [264]. Similarly, Chapman-Shimshoni et al [265] showed that exposure of clonal B lymphocytes from patients with Chronic lymphocytic leukemia to simvastatin decreases viability significantly by the induction of apoptosis. The apoptosis induced by simvastatin was probably initiated by the mitochondrial caspase 9, which indirectly leads to activation of caspase 3 and 8.

On the other hand, Rezaie-Majd et al [266] showed that 6 weeks of *in vivo* statin treatment reduced the binding of peripheral blood mononuclear cells (PBMCs) to TNF-α-stimulated HUVECs *in vitro* due to significant reduction of CD54, CD18 and CD11a-mRNA levels in PBMCs compared with their respective pre-treatment levels. Similary, Grip et al [267], showed that treatment of human peripheral monocytes with atorvastatin for up to 24 h activated PPAR-γ with inhibition of the production of TNF-α, MCP-1 and MMP-9. Recently, Frigerio et al [268], demonstrated inhibition of transmigration of PBMCs through endothelium by *in vivo* statin treatment in humans. In healthy control donors and in hypercholesterolaemic patients transmigration of PBMCs through unstimulated cells was significantly lower than transmigration observed across TNF-α-stimulated endothelium at all time points studied.

In summary, all the above-described anti-inflammatory effects of statins can either result from their capacity to interfere with the mevalonate pathway (Figure 2) and inhibit prenylation of Rho family GTPases, or due to an effect independent of the mevalonate pathway. Statins through those effects have immunomodulatory activities, including suppression of T-cell activation, anti-inflammatory activities both on macrophages and neutrophils and inhibition of several proinflammatory cytokines, including TNF-α, IL-6 and IL-8. On the other hand, it has been demonstrated that inhibition of MAPK pathways, including p38 MAPK, ERK-1/2, and JNK in different cells from RA patients, prevented the expression of RANKL, the production of MMPs and cytokines, and cartilage destruction [for review, see ref. 269].

For all those reasons, the statins through their immunomodulatory and anti-inflammatory properties may be the next step in the treatment of SLE patients (Figure 5).

References

[1] Goldstein LG, Brown MS. Regulation of the mevalonate pathway. *Nature* 1990;343:425-430.

[2] Laufs U, Fata VL, Liao JK. Inhibition of 3-hydroxy-3-methylglutaryl (HMG)-CoA reductase blocks hipoxia-mediated down-regulation of endothelial nitric oxid synthase. *J Biol Chem* 1997;272:31725-31729.

[3] Hernandez-Perera O, Perez-Sala D, Navarro-Antolin J, Sanchez-Pascuala R, Hernandez G, Diaz C, Lamas S. Effects of 3-hydroxy-3-methylglutaryl-CoA reductase inhibitors,

artovastatin and simvastatin, on the expression of endothelin-1 and endothelial nitric oxide synthase in vascular endothelial cells. *J Clin Invest*. 1998;101:2711-2719.

[4] Zhang FL, Casey PJ. Protein prenylation: molecular mechanisms and functional consequences. *Annu Rev Biochem*. 1996;65:241-269.

[5] Macara IG, Lounsbury KM, Richards SA, McKiernan C, Bar-Sagi D. The Ras superfamily of GTPases. *FASEB J*. 1996;10:625-630.

[6] Mackay DJ, Hall A. Rho GTPases. *J Biol Chem*. 1998;273:20685-20688.

[7] Somsel Rodman J, Wandinger-Ness A. Rab GTPases coordinate endocytosis. *J Cell Sci*. 2000;113 Pt 2:183-192.

[8] Robinson MJ, Cobb MH. Mitogen-activated kinase pathways. *Curr Opin Cell Biol* 1997;9:180-186.

[9] Zarubin T, Han J. Activation and signaling of the p38 MAP kinase pathway. *Cell Res* 2005;15(1):11-18.

[10] Jiang Y, Chen C, Li Z, Guo W, Gegner JA, Lin S, Han J. Characterization of the structure and function of a new mitogen-activated protein kinase (p38β). *J Biol Chem* 1996; 271(30):17920-17926.

[11] Lechner C, Zahalka MA, Giot JF, Moler NPH, Ullrich A. ERK6, a mitogen-activated protein kinase involved in C2C12 myoblast differentiation. *Proc Natl Acad Sci USA* 1996; 93:4355-4359.

[12] Li Z, Jiang Y, Ulevitch RJ, Han J. The primary structure of p38gamma: A new member of p38 group of MAP kinase. *Biochem Biophys Res Comm* 1996; 228:334-340.

[13] Jiang Y, Gram H, Zhao M, New L, Gu J, Feng L, Di Padova F, Ulevitch RJ, Han J. Characterization of the structure and function of the fourth member of p38 group mitogenactivated protein kinases, p38delta. *J Biol Chem* 1997; 272:30122-30128.

[14] Kumar S, McDonnell PC, Gum RJ, Hand AT, Lee JC, Young PR. Novel homologues of CSBP/p38 MAP kinase: activation, substrate specificity and sensitivity to inhibition by pyridinyl imidazoles. *Biochem Biophys Res Commun* 1997; 235:533-538.

[15] Mii S, Khalil RA, Morgan KG, Ware JA, Kent KG. Mitogen-activated protein kinase and proliferation of human vascular smooth muscle cells. *Am J Physiol* 1996;270:H142-H150.

[16] Force T, Bonventre JV. Growth factors and mitogen-activated protein kinases. *Hypertension* 1998;31:152-161.

[17] Touyz RM, Schiffrin EL. Ang II-stimulated generation of reactive oxygen species in human vascular smooth muscle cells is mediated via phospholipase D-dependent pathways. *Hypertension* 1999;34:976-982.

[18] Ichijo H. From receptors to stress-activated MAP kinases. *Oncogene* 1999;18:6087-6093.

[19] Morinville A, Maysinger D, Shaver A. From vanadis to atorpos:vanadium compounds as pharmacological tools in cell death signaling. *Trends Pharmacol Sci* 1998;19:452–459.

[20] New L, Han J. The p38 MAP Kinase pathway. *Trends Cardiovasc Med* 1998;8:220–229.

[21] Ip YT, Davis RJ. Signal transduction by the c-Jun N-terminal kinase (JNK)-from inflammation to development. *Curr Opin Cell Biol* 1998;10:205–219.

[22] Blanco-Colio LM, Tunon J, Martin-Ventura JL, Egido J. Anti-inflammatory and immunomodulatory effects of statins. *Kidney Int* 2003;63:12–23.

[23] van Seventer GA, Semnani RT, Palmer EM, McRae BL, van Seventer JM. Integrins and T helper cell activation. *Transplant Proc* 1998;30:4270–4.

[24] Weitz-Schmidt G, Welzenbach K, Brinkmann V, Kamata T, Kallen J, Bruns C, Cottens S, Takada Y, Hommel U. Statins selectively inhibit leukocyte function antigen-1 by binding to a novel regulatory integrin site. *Nat Med* 2001;7:687–92.

[25] Youssef S, Stuve O, Patarroyo JC, Ruiz PJ, Radosevich JL, Hur EM, Bravo M, Mtchell DJ, Sobel RA, Steinman L, Zamvil SS. The HMG-CoA reductase inhibitor, atorvastatin, promotes a Th2 bias and reverses paralysis in central nervous system autoimmune disease. *Nature* 2002;420:78–84.

[26] Cantrell DA. GTPases and T cell activation. *Immunol Rev* 2003;192:122–30.

[27] Ghittoni R, Patrussi L, Pirozzi K, Pellegrini M, Lazzerini PE, Capecchi PL, Pasini FL, Baldari CT. Simvastatin inhibits T-cell activation by selectively impairing the function of Ras superfamily GTPases. *FASEB J* 2005; 19:605–7.

[28] Blake GJ, Ridker PM. Are statins anti-inflammatory? *Curr Control Trials Cardiovasc Med* 2000;1:161–5.

[29] Weitz-Schmidt G. Statins as anti-inflammatory agents. *Trends Pharmacol Sci* 2002;23:482–6.

[30] Kwak B, Mulhaupt F, Myit S, Mach F. Statins as a newly recognized type of immunomodulator. *Nat Med* 2000;6:1399–402.

[31] Essig M, Nguyen G, Prie D, Escoubet B, Sraer JD, Friedlander G. 3-Hydroxy-3-methylglutaryl coenzyme A reductase inhibitors increase fibrinolytic activity in rat aortic endothelial cells. Role of geranylgeranylation and Rho proteins. *Circ Res* 1998;83:683–90.

[32] Ortego M, Bustos C, Hernandez-Presa MA, Tunon J, Diaz C, Hernandez G, Egido J. Atorvastatin reduces NF-kappaB activation and chemokine expression in vascular smooth muscle cells and mononuclear cells. *Atherosclerosis* 1999;147:253–61.

[33] Hausding M, Witteck A, Rodriguez-Pascual F, von Eichel-Streiber C, Forstermann U, Kleinert H. Inhibition of small G proteins of the rho family by statins or clostridium difficile toxin B enhances cytokine-mediated induction of NO synthase II. *Br J Pharmacol* 2000;131:553–61.

[34] Danesh FR, Sadeghi MM, Amro N, Philips C, Zeng L, Lin S, Sahai A, Kanwar YS. 3-Hydroxy-3-methylglutaryl CoA reductase inhibitors prevent high glucose-induced proliferation of mesangial cells via modulation of Rho GTPase/p21 signaling pathway: implications for diabetic nephropathy. *Proc Natl Acad Sci U S A* 2002;99:8301–5.

[35] Eto M, Kozai T, Cosentino F, Joch H, Luscher TF. Statin prevents tissue factor expression in human endothelial cells: role of Rho/Rho-kinase and Akt pathways. *Circulation* 2002;105:1756–9.

[36] Masamura K, Oida K, Kanehara H, Suzuki J, Horie S, Ishii H, Miyamori I. Pitavastatin-induced thrombomodulin expression by endothelial cells acts via inhibition of small G proteins of the Rho family. *Arterioscler Thromb Vasc Biol* 2003;23:512–7.

[37] Trager J, Ward MM. Mortality and causes of death in systemic lupus erythematosus. *Curr Opin Rheumatol* 2001;13:345–51.

[38] Sturfelt G, Eskilsson J, Nived O, Truedsson L, Valind S. Cardiovascular disease in systemic lupus erythematosus: a study of 75 patients from a defined population. *Medicine* (Baltimore) 1992;71:216–23.

[39] Ward MM. Premature morbidity from cardiovascular and cerebrovascular diseases in women with systemic lupus erythematosus. *Arthritis Rheum* 1999;42:338–46.

[40] Shome GP, Sakauchi M, Yamane K, Takemura H, Kashiwagi H. Ischemic heart disease in systemic lupus erythematosus: a retrospective study of 65 patients treated with prednisolone. *Jpn J Med* 1989;28:599–603.

[41] Esdaile JM, Abrahamowicz M, Grodzicky T, Senecal JL, Panaritis T, Li Y, et al. Myocardial infarction (MI) and stroke in SLE: markedly increased incidence after controlling for risk factors. *Arthritis Rheum* 1998;41:S139.

[42] Manzi S, Meilahn EN, Rairie JE, Conte CG, Medsger TA Jr, Jansen-McWilliams L, DÁgostino RB, Kuller LH. Age-specific incidence rates of myocardial infarction and angina in women with systemic lupus erythematosus: comparison with the Framingham Study. *Am J Epidemiol* 1997;145:408–15.

[43] Esdaile JM, Abrahamowicz M, Grodzicky T, Li Y, Panaratis C, du Berger R, Cote R, Grover SA, Fortin PR, Clarke AE, Senecal JL. Traditional Framingham risk factors fail to fully account for accelerated atherosclerosis in systemic lupus erythematosus. *Arthritis Rheum* 2001;44:2331–7.

[44] Shepherd J, Cobbe SM, Ford I, Isles CG, Lorimer AR, MacFarlane PW, McKillop JH, Packard CJ. Prevention of coronary heart disease with pravastatin in men with hypercholesterolemia. West of Scotland Coronary Prevention Study Group. *N Engl J Med.* 1995;333:1301-1307.

[45] Downs JR, Clearfield M, Weis S, Whitney E, Shapiro DR, Beer PA, Langendorfer A, Stein EA, Kruyer W, Goto AM Jr. Primary prevention of acute coronary events with lovastatin in men and women with average cholesterol levels: results of AFCAPS/TexCAPS. Air Force/Texas Coronary Atherosclerosis Prevention Study. *JAMA.* 1998;279:1615-1622.

[46] Scandinavian Simvastatin Survival Study Group. Randomised trial of cholesterol lowering in 4444 patients with coronary heart disease: the Scandinavian Simvastatin Survival Study (4S). *Lancet.* 1994;344:1383-1389.

[47] Sacks FM, Pfeffer MA, Moye LA, Rouleau JL, Rutherford JD, Cole TG, Brown L, Warnica JW, Arnold JM, Wun CC, Davis BR, Braunwald E. The effect of pravastatin on coronary events after myocardial infarction in patients with average cholesterol levels. Cholesterol and Recurrent Events (CARE) Trial Investigators. *N Engl J Med.* 1996;335:1001-1009.

[48] Prevention of cardiovascular events and death with pravastatin in patients with coronary heart disease and a broad range of initial cholesterol levels. Long-Term Intervention with Pravastatin in Ischaemic Disease (LIPID) Study Group. *N Engl J Med.*1998;339:1349-1357.

[49] Schwartz GG, Olsson AG, Ezekowitz MD, Ganz P, Oliver MF, Waters D, Zeiher A, Chaitman BR, Leslie S, Stern T. Myocardial Ischemia Reduction with Aggressive Cholesterol Lowering (MIRACL) Study Investigators. Effects of atorvastatin on early

recurrent ischemic events in acute coronary syndromes: the MIRACL study: a randomized controlled trial. *JAMA*. 2001;285:1711-1718.

[50] Laufs U, Fata VL, Liao JK. Inhibition of 3-hydroxy-3-methylglutaryl (HMG)-CoA reductase blocks hipoxia-mediated down-regulation of endothelial nitric oxid synthase. *J Biol Chem* 1997;272(50):31725-31729.

[51] Hernandez-Perera O, Perez-Sala D, Navarro-Antolin J, Sanchez-Pascuala R, Hernandez G, Diaz C, Lamas S. Effects of 3-hydroxy-3-methylglutaryl-CoA reductase inhibitors, artovastatin and simvastatin, on the expression of endothelin-1 and endothelial nitric oxide synthase in vascular endothelial cells. *J Clin Invest* 1998;101(12):2711-2719.

[52] English J, Pearson G, Wilsbacher J, Swantek J, Karandikar M, Xu S, Cobb MH. New insights into the control of MAP kinase pathways. *Exp Cell Res*. 1999;253:255-270.

[53] Ardaillou R. Angiotensin II receptors. *J Am Soc Nephrol*. 1999;10:S30-S39.

[54] de Gasparo M, Husain A, Alexander W, Catt KJ, Chiu AT, Drew M, Goodfriend T, Harding JW, Inagami T, Timmermans BMWM. Proposed update of angiotensin receptor nomenclature. *Hypertension* 1995;25:924-927.

[55] Errede B, Levin DE. A conserved kinase cascade for MAP kinase activation in yeast. *Curr Opin Cell Biol*. 1993;5(2):254-60.

[56] Davis RJ. MAPKs: new JNK expands the group. *Trends Biochem Sci*. 1994;19(11):470-3.

[57] Waskiewicz AJ and Cooper JA. Mitogen and stress response pathways: MAP kinases cascades and phosphatase regulation in mammals and yeast. *Curr Opin Cell Biol*. 1995;7(6):798-805.

[58] Kyriakis JM and Avruch J. Sounding the alarm: protein kinase cascades activated by stress and inflammation. *J Biol Chem*. 1996;271(40):24313-6.

[59] Widmann C, Gibson S, Jarpe MB, Johnson GL. Mitogen-activated protein kinases: conservation of a three-kinase module from yeast to human. *Physiol Rev*. 1999;79(1):143-80.

[60] Schaeffer HJ, Weber MJ. Mitogen-activated protein kinases: specific messages from ubiquitous messengers. *Mol Cell Biol*. 1999;19(4):2435-44.

[61] Cobb MH, Goldsmith EJ. How MAP kinases are regulated. *J Biol Chem*. 1995;270(25):14843-6.

[62] Kyriaskis JM, Avruch J. pp54 microtubule-associated protein 2 kinase. A novel serine/threonine protein kinase regulated by phosphorylation and stimulated by poly-L-lysine. *J Biol Chem* 1990;265:17355-17363.

[63] Hibi M, Lin A, Smeal T, Minden A, Karin M. Identification of an oncoprotein- and UV-responsive protein kinase that binds and potentiates the c-Jun activation domain. *Genes Dev*. 1993;7:2135-2148.

[64] Han J, Lee JD, Bibbs L, Ulevitch RJ. A MAP kinase targeted by endotoxin and hyperosmolarity in mammalian cells. *Science* 1994;265:808-811.

[65] Lee JC, Laydon JT, Mcdonnell PC, Gallagher TF, Kumar S, Green D, McNulty D, Blumenthal MJ, Heys JR, Landvatter SW. A protein kinase involved in the regulation of inflammatory cytokine biosynthesis. *Nature* 1994;372:739-746.

[66] Zhou G, Bao ZQ, Dixon JE. Components of a new human protein kinase signal transduction pathway. *J Biol Chem* 1995;270:12665-12669.

[67] Lee JD, Ulveitch RJ, Han J. Primary structure of BMK1: A new mammalian MAP kinase. *Biochem Biophys Res Commun* 1995;213:715-724.

[68] Ushio-Fukai M, Alexander RW, Akers M, Griendling KK. p38 MAP kinase is a critical component of the redox-sensitive signaling pathways by angiotensin II: role in vascular smooth muscle cell hypertrophy. *J Biol Chem*. 1998;273:15022–15029.

[69] Yoshizumi M, Abe J, Haendeler J, Huang Q, Berk BC. Src and Cas mediate JNK activation but not ERK1/2 and p38 kinases by reactive oxygen species. *J Biol Chem*. 2000;275:11706–11712.

[70] Zhang J, Jin N, Liu Y, Rhoades RA. Hydrogen peroxide stimulates extracellular signal-regulated protein kinases in pulmonary arterial smooth muscle cells. *Am J Respir Cell Mol Biol*. 1998;19:324–332.

[71] Baas AS, Berk BC. Differential activation of mitogen-activated protein kinases by H_2O_2 and O_2 in vascular smooth muscle cells. *Circ Res*. 1995;77:29–36.

[72] Viedt C, Soto U, Krieger-Brauer HI, Fei J, Elsing C, Kubler W, Kreuzer J. Differential activation of mitogen-activated protein kinases in smooth muscle cells by angiotensin II: involvement of p22phox and reactive oxygen species. *Arterioscler Thromb Vasc Biol*. 2000;20:940–948.

[73] Shin EA, Kim KH, Han SI, Ha KS, Kim JH, Kang KI, Kim HD, Kang HS. Arachidonic acid induces the activation of the stress-activated protein kinase, membrane ruffling and H2O2 production via a small GTPase Rac1. *FEBS Lett*. 1999;452:355–359.

[74] Sundaresan M, Zu-Xi Y, Ferrans VJ, Irani K, Finkel T. Requirement for generation of H_2O_2 for platelet-derived growth factor signal transduction. *Science*. 1995;270:296–299.

[75] Leduc I, Meloche S. Angiotensin II stimulates tyrosine phosphorylation of the focal adhesion protein paxillin in aortic smooth mucle cells. *J Biol Chem* 1995;270:4401–4404.

[76] Kudoh S, Komuro I, Mizuno T, Yamazaki T, Zou Y, Shiojima I, Takekoshi N, Yazaki Y. Angiotensin II stimulates c-Jun NH2-terminal kinase in culturedcardiac myocytes of neonatal rats. *Circ Res* 1997;80:139–146.

[77] Touyz RM, El Mabrouk M and Schiffrin EL. MEK inhibition attenuates Ang II-mediated signaling and contraction in SHR vascular smooth muscle cells. *Circ Res* 1999;84:505–515.

[78] Ushio-Fukai M, Alexander RW, Akers M, Griendling KK. p38 Mitogen-activated protein kinase is a critical component of the redoxsensitive signaling pathways activated by angiotensin II. Role in vascular smooth muscle cell hypertrophy. *J Biol Chem* 1998;273:15022–15029.

[79] Ushio-Fukai M, Alexander RW, Akers M, Yin Q, Fujio Y, Walsh K, Griendling KK. Reactive oxygen species mediate the activation of Akt/protein kinase B by angiotensin II in vascular smooth muscle cells. *J Biol Chem* 1999;274:22699–22704.

[80] Hamaguchi A, Kim S, Yano M, Yamanaka S and Iwao H. Activation of glomerular mitogen-activated protein kinases in angiotensin II-mediated hypertension. *J Am Soc Nephrol* 1998;9:372 – 380.

[81] Izumi Y, Kim S, Murakami T, Yamanaka S and Iwao H. Cardiac mitogen-activated protein kinase activities are chronically increased in stroke-prone hypertensive rats. *Hypertension* 1998;31:50 – 56.

[82] Kim S, Murakami T, Izumi Y, Yano M, Miura K, Yamanaka S, Iwao H: Extracellular signal-regulated kinase and c-Jun NH2-terminal kinase activities are continuously and differentially increased in aorta of hypertensive rats. *Biochem Biophys Res Commun* 1997;236:199 – 204.

[83] Kim S, Izumi Y, Yano M, Hamaguchi A, Miura K, Yamanaka S, Miyazaki H, Iwao H. Angiotensin blockade inhibits activation of mitogen-activated protein kinases in rat ballooninjured artery. *Circulation* 1998;97:1731 – 1737.

[84] Kim S, Zhan Y, Izumi Y, Yasumoto H, Yano M, Iwao H. In vivo activation of rat aortic platelet-derived growth factor and epidermal growth factor receptors by angiotensin II and hypertension. *Arterioscler Thromb Vasc Biol* 2000;20:2539 – 2545.

[85] Kim S, Iwao H. Molecular and cellular mechanisms of angiotensin II-mediated cardiovascular and renal diseases. *Pharmacol Rev* 2000;52:11 – 34.

[86] Yano M, Kim S, Izumi Y, Yamanaka S, Iwao H. Differential activation of cardiac c-Jun amino-terminal kinase and extracellular signal-regulated kinase in angiotensin II-mediated hypertension. *Circ Res* 1998;83:752 – 760.

[87] Dong C, Davis RJ, Flavell RA. MAP kinases in the immune response. *Annu Rev Immunol* 2002;20:55–72.

[88] Pietersma A, Tilly BC, Gaestel M, de Jong N, Lee JC, Koster JF, Sluiter W. p38 mitogen activated protein kinase regulates endothelial VCAM-1 expression at the post-transcriptional level. *Biochem Biophys Res Commun* 1997;230(1):44–8.

[89] Guan Z, Buckman SY, Pentland AP, Templeton DJ, Morrison AR. Induction of cyclooxygenase-2 by the activated MEKK1→ SEK1/MKK4 → p38 mitogen-activated protein kinase pathway. *J Biol Chem* 1998;273(21):12901–8.

[90] Badger AM, Cook MN, Lark MW, Newman-Tarr TM, Swift BA, Nelson AH, Barone FC, Kumar S. SB 203580 inhibits p38 mitogenactivated protein kinase, nitric oxide production, and inducible nitric oxide synthase in bovine cartilage-derived chondrocytes. *J Immunol* 1998;161(1):467–73 [Erratum in: *J Immunol* 1999;162(5):3105].

[91] Da Silva J, Pierrat B, Mary JL, Lesslauer W. Blockade of p38 mitogen-activated protein kinase pathway inhibits inducible nitric-oxide synthase expression in mouse astrocytes. *J Biol Chem* 1997;272(45):28373–80.

[92] Rincon M, Whitmarsh A, Yang DD, Weiss L, Derijard B, Jayaraj P, Davis RJ, Flavell RA. The JNK pathway regulates the In vivo deletion of immature CD4(+)CD8(+) thymocytes. *J Exp Med* 1998;188 (10): 1817–30.

[93] Yang DD, Conze D, Whitmarsh AJ, Barrett T, Davis RJ, Rincon M, Flavell RA. Differentiation of CD4+ T cells to Th1 cells requires MAP kinase JNK2. *Immunity* 1998;9(4):575–85.

[94] Faris M, Kokot N, Lee L, Nel AE. Regulation of interleukin-2 transcription by inducible stable expression of dominant negative and dominant active mitogen-activated protein kinase kinase kinase in jurkat T cells. Evidence for the importance of

Ras in a pathway that is controlled by dual receptor stimulation. *J Biol Chem* 1996;271(44):27366–73.

[95] Dong C, Yang DD, Tournier C, Whitmarsh AJ, Xu J, Davis RJ, Flavell RA. JNK is required for effector T-cell function but not for Tcell activation. *Nature* 2000;405(6782):91–4.

[96] Skov S. Intracellular signal transduction mediated by ligation of MHC class I molecules. *Tissue Antigens* 1998;51(3):215–23.

[97] Smith A, Ramos-Morales F, Ashworth A, Collins M. A role for JNK/SAPK in proliferation, but not apoptosis, of IL-3-dependent cells. *Curr Biol* 1997;7(11):893–6.

[98] Tuyt LM, Dokter WH, Birkenkamp K, Koopmans SB, Lummen C, Kruijer W, Vellenga E. Extracellular-regulated kinase 1/2, Jun Nterminal kinase, and c-Jun are involved in NF-kappa Bdependent IL-6 expression in human monocytes. *J Immunol* 1999;162(8): 4893–9402.

[99] Min W, Pober JS. TNF initiates E-selectin transcription in human endothelial cells through parallel TRAF-NF-kappa B and TRAF-RAC/CDC42-JNK-c-Jun/ATF2 pathways. *J Immunol* 1997;159(7):3508–18.

[100] Blasini AM, Alonzo E, Chacon R, Riera R, Stekman IL, Rodriguez MA. Abnormal pattern of tyrosine phosphorylation in unstimulated peripheral blood. T lymphocytes from patients with systemic lupus erythematosus, *Lupus* 1998;7:1–9.

[101] Blasini AM, Brundula V, Paris M, Rivas L, Salazar S, Stekman IL, Rodriguez MA. Protein tyrosine kinase (PTK) activity in T lymphocytes from patients with systemic lupus erithematosus (SLE), *J. Autoimmun*. 1998;11:387–393.

[102] Liossis SNC, Ding XZ, Dennis GJ, Tsokos GC. Altered pattern of TCR/CD3-mediated protein-tyrosil phosphorylation in T cells from patients with systemic lupus erythematosus. Deficient expression of the T cell receptor zeta chain, *J. Clin. Invest.* 1998;101:1448–1457.

[103] Matache C, Stefanescu M, Onu A, Szegli G, Barel M, Tanaseanu S, Matei I, Bouillie S, Frade R, Tyrosine phosphorylation in peripheral blood lymphocytes from patients with systemic lupus erythematosus, *Autoimmunity* 1996;24:217–228.

[104] Brundula V, Rivas L, Blasini AM, Paris M, Salazar S, Stekman IL, Rodriguez MA. Diminished levels of TCR ξ chains in peripheral blood lymphocytes from patients with systemic lupus erythematosus, *Arthritis Rheum*. 1999;42:1908–1916.

[105] Takeuchi T, Tsukada K, Pang M, Amano K, Koide J, Abe T. TCR ξ chain lacking exon 7 in two patients with systemic lupus erythematosus, *Int. Immunol*. 1998;10:911–921.

[106] Cedeno S, Cifarelli DF, Blasini AM, Paris M, Placeres F, Alonso G, Rodriguez MA. Defective activity of ERK-1 and ERK-2 mitogen-activated protein kinases in peripheral blood T lymphocytes from patients with systemic lupus erythematosus: potential role of altered coupling of Ras guanine nucleotide exchange factor hSos to adapter protein Grb2 in lupus T cells. *Clin Immunol*. 2003;106(1):41-9.

[107] Deng C, Kaplan MJ, Yang J, Ray D, Zhang Z, McCune WJ, Hanash SM, Richardson BC. Decreased Ras-mitogen-activated protein kinase signaling may cause DNA hypomethylation in T lymphocytes from lupus patients. *Arthritis Rheum*. 2001;44: 397–407

[108] Yi Y, McNerney M, Datta SK. Regulatory defects in Cbl and mitogen-activated protein kinase (extracellular signal-related kinase) pathways cause persistent hyperexpression of CD40 ligand in human lupus T cells. *J. Immunol.* 2000;165: 6627–6634

[109] Rapoport MJ, Amit M, Aharoni D, Weiss M, Weissgarten J, Bruck N, Buchs A, Bistritzer T, Molad Y. Constitutive up-regulated activity of MAP kinase is associated with down-regulated early p21Ras pathway in lymphocytes of SLE patients. *J Autoimmun.* 2002;19(1-2):63-70.

[110] Wei L, Wang J, Thompson P, Wong S, Struthers AD, MacDonald TM. Adherence to statin treatment and readmission of patients after myocardial infarctation: a six-year follow u study. *Heart.* 2002;88(3):229-233.

[111] 2001 Heart and Stroke Statistical Update [cited 2001 Mar 11]. Available from: URL: http://www.americanheart.org/statistics/coronary.html.

[112] Castelli WP. Epidemiology of coronary heart disease: Framingham study. *Am J Med* 1984;76(suppl):4-12.

[113] Chong PH, Bachenheimer BS. Current, new, and future treatments in dyslipidaemia and atherosclerosis. *Drugs* 2000;60:55-93.

[114] Shepherd J, Cobbe SM, Ford I, Isles CG, Lorimer AR, MacFarlane PW McKillop JH, Pakard CJ. for the West of Scotland Coronary Prevention Study (WOSCOPS) Group. Prevention of coronary heart disease with pravastatin in men with hypercholesterolemia. *Atheroscler Suppl.* 2004;5(3):91-7.

[115] Massy ZA, Keane WF, Kasiske BL. Inhibition of the mevalonate pathway: benefits beyond cholesterol reduction? *Lancet* 1996;347(8994):102-103.

[116] Blum CB. Comparison of properties of four inhibitors of 3-hydroxy-3-methylglutaryl-coenzyme A reductase. *Am J Cardiol* 1994;73(14):3D-11D.

[117] Voughan CJ, Murphy MB, Buckley BM. Statins do more than just lower cholesterol. *Lancet* 1996;348:1079-1082.

[118] Endres M, Laufs U, Huang Z, Nakamura T, Huang P, Moskowitz MA, Liao JK. Stroke protection by 3-hydroxy-3-methylglutaryl (HMG)-CoA reductase inhibitors mediated by endothelial nitric oxide synthase. *Proc Natl Acad Sci* 1998;95(15):8880-8885.

[119] Rosenson RS, Tangney CC. Antiatherothrombotic properties of statins: implications for cardiovascular event reduction. *JAMA* 1998;279(20):1643-1650.

[120] Laufs U, Liao JK. Targeting Rho in cardiovascular disease. *Circ Res* 2000;87(7):526-528.

[121] Ichiki T, takeda K, Tokunou T, iino N, Egashira K, Shimokawa H, Hirano K, Kanaide H, Takeshita A. Down regulation of angiotensin II type 1 receptor blocker, valsartan, on 3-hydroxy-3-methylglutaryl-coenzyme A reductase inhibitors in vascular smooth muscle cells. *Arterioscler Thromb Vasc Biol* 2001;21(12):1896-1901.

[122] van Haelst PL, van Doormaal JJ, May JF, Gans RO, Crijns HJ, Cohen Tervaert JW. Secondary prevention with fluvastatin decreases levels of adhesion molecules, neopterin and C-reactive protein. *Eur J Intern Med.* 2001;2(6):503-509.

[123] Wassmann S, Laufs U, Baumer AT, Muller K, Konkol C, Sauer H, Bohm M, Nickening G. Inhibition of geranylgeranylation reduces angiotensin II-mediated free radical production in vascular smooth muscle cells: involment of angiotensin AT1 receptor expression and Rac1 GTPase. *Mol Pharmacol.* 2001;59(3):646-654.

[124] Kalinowski L, Dobrucki LW, Brovkovych V, Malinski T. Increased nitric oxide bioavailability in endothelial cells contributes to the pleiotropic effect of cerivastatin. *Circulation* 2002;105(8):933-938.

[125] Szczeklik A, undas A, Musial J, Gajewski P, Swadzba J, Jankowski M. Antithrombotic actions of statins. *Med Sci Monit*. 2001;7(6):1381-1385.

[126] Chung HK, Lee IK, Kang H, Suh JM, Kim H, Park KC, Kim DW, Kim YK, Ro HK, Shong M. Statin inhibets interferon-gammag-induceexpression of intracellular adhesion molecule-1 (ICAM-1) in vascular endothelial and smooth muscle cells. *Exp Mol Med.* 2002;34(6):451-461.

[127] Laufs U, Gertz K, dirnagl U, Bohm M, Nickenig G, Endres M. Rosuvastatin, a new HMG-CoA reductase inhibitor, upregulates endothelial nitric oxide synthase and protects from ischemic stroke in mice. *Brain Res*. 2002;942(1-2):23-30.

[128] Seljeflot I, Tonstad S, Hjermann I, Arnesen H. Improved fibrinolysis after 1-year treatment with HMG-CoA reductase inhibitors in patients with coronary heart disease. *Thromb Res* 2002;105(4):285-290.

[129] Venugopal SK, Devaraj S, Yuhanna I, Shaul P, Jialal I. Demonstration that C-reactive protein decrease eNOS expression and bioactivity in human aortic endothelial cells. *Circulation* 2002;106:1439-1441.

[130] Greenwood J, Mason JC. Statins and the vascular endothelial inflammatory response. *Trends Immunol*. 2007;28(2):88-98.

[131] Touys RM, Schiffrin E. Signal transduction mechanisms mediating the physiological and pathophysiological actions of angiotensin II in vascular smooth muscle cells. *Pharmacol Rev*. 2000;52(4):639-672.

[132] Denhardt DT. Signal-transducing protein phosphorylation cascades mediated by Ras/Rho proteins in the mammalian cell: the potential for multiplex signalling. *Biochem J*. 1996;318 (Pt 3):729-47.

[133] Liao DF, Monia B, Dean N, Berk BC. Protem kmase C-ξ mediates angotensin II activation of ERKl/2 in vascular smooth muscle cells. *J Biol Chem* 1997;272: 6146-6150.

[134] Kyriakis JM, Force TL, Rapp UR, Bonventre JV, Avruch J. Mitogen regulation of c-Raf protein kinase activity toward miltogen-activated protein kinase kinase. *J Biol Chem* 1993;268:16009-16019.

[135] Wang Y, Sunonson MS, Pouyssegur J, Dunn MJ. Endothelin rapidly stimulates mitogen-activated protein kinase activity in rat mesangial cells. *Biochem J* 1992;287: 589-594.

[136] Wang Y, Rose PM, Webb ML, Dunn MJ. Endothelin stimulates mitogen-activated protein kinase cascade through either ETA or ETB. *Am J Phisiol* 1994;267: Cl130-Cl135.

[137] Monnelli TA, Zhang L-M, Newman WH, Meier KE. Thromboxane A2/prostaglandin H2-stimulated mitogenesis of coronary artery smooth muscle cells involves activation of mitogen-activated protein kinase and S6 kinase. *J Biol Chem* 1994 ;269:5693-5698.

[138] Watanabe T, Waga I, Honda Z, Kurokawa K, Shimizu T. Prostaglandin $F_{2\alpha}$ stimulates formation of $p21^{ras}$-GTP complex and mitogen-activated protein kinase in NIH-3T3 cells via Gq-protein-coupled pathway. *J Biol Chem* 1995 ;270 : 8984-8990.

[139] Muthahf MM, Benter IF, Uddin MR, Mahk KU. Calcium/calmodulin- dependent protein kinase Iia mediates activation of mitogen-activated protein kinase and cytosolic phospholipase A2 in norepinephrine-induced arachidonic acid release in rabbit aortic smooth muscle cells. *J Biol Chem* 1996;271:30149-30157.

[140] Wan Y, Kurosaki T, Huang X. Tyrosine kinases in activation of the MAP kinase cascade by G-protein-coupled receptors. *Nature* 1996;380:541-544.

[141] Reynolds CM, Eguchi S, Frank GD, Motley ED. Signaling mechanisms of heparin-binding epidermal growth factor-like growth factor in vascular smooth muscle cells. *Hypertension*. 2002;39(2 Pt 2):525-9.

[142] Murasawa S, Mori Y, Nozawa Y, Gotoh N, Shibuya M, Masaki H, Maruyama K, Tsutsumi Y, Moriguchi Y, Shibazaki Y, Tanaka Y, Iwasaka T, Inada M, Matsubara H. Angiotensin II type 1 receptor–induced extracellular signal–regulated protein kinase activation is mediated by Ca^{2+}/calmodulin-dependent transactivation of epidermal growth factor receptor. *Circ Res*. 1998; 82: 1338–1348.

[143] Eguchi S, Numaguchi K, Iwasaki H, Matsumoto T, Yamakawa T, Utsunomiya H, Motley ED, Kawakatsu H, Owada KM, Hirata Y, Marumo F, Inagami T. Calcium-dependent epidermal growth factor receptor transactivation mediates the angiotensin II-induced mitogen-activated protein kinase activation in vascular smooth muscle cells. *J Biol Chem*. 1998; 273: 8890–8896.

[144] Saito Y, Berk BC. Transactivation: a novel signaling pathway from angiotensin II to tyrosine kinase receptors. *J Mol Cell Cardiol*. 2001;33(1):3-7.

[145] Prenzel N, Zwick E, Daub H, Leserer M, Abraham R, Wallasch C, Ullrich A. EGF receptor transactivation by G-protein-coupled receptors requires metalloproteinase cleavage of proHB-EGF. *Nature* 1999;402(6764):884-888.

[146] Hyder SM, Stancel GM, Chiappetta C, Murthy L, Boettger-Tong HL, Makela S. Uterine expression of vascular endothelial growth factor is increased by estradiol and tamoxifen. *Cancer Res*. 1996; 56:3954-3960.

[147] Mustonen T, Alitalo K. Endothelial receptor tyrosine kinases involved in angiogenesis. *J. Cell Biol*. 1995:129:895 -898.

[148] Soker S, Fidder H, Neufeld G, Klagsbrun M. Characterization of novel vascular endothelial growth factor (VEGF) receptors on tumor cells that bind VEGF165 via its exon 7-encoded domain. *J.Biol. Chem*. 1996;271:5761-5767.

[149] Guo D, Jia Q, Song HY,Warren RS, Donner DB. Vascular endothelial cell growth factor promotes tyrosine phosphorylation of mediators of signal transduction that contain SH2 domains. Association with endothelial cell proliferation. *J. Biol. Chem*. 1995;270: 6729-6733.

[150] Seetharam L, Gotoh N, Maru Y, Neufeld G, Yamaguchi S, Shibuya M. A unique signal transduction from FLT tyrosine kinase, a receptor for vascular endothelial growth factor VEGF. *Oncogene* 1995; 10:135-147.

[151] Cunningham SA, Waxham MN, Arrate PM and Brock TA. Interaction of the Flt-1 tyrosine kinase receptor with the p85 subunit of phosphatidylinositol 3-kinase. Mapping of a novel site involved in binding. *J. Biol. Chem*.1995:270:20254-20257.

[152] Breier G and Risau W. The role of vascular endothelial growth factor in blood vessel formation. *Trends Cell Biol*. 1996:6:454-456.

[153] D'Angelo G, Struman I, Martial J, Weiner RI. Activation of mitogen-activated protein kinases by vascular endothelial growth factor and basic fibroblast growth factor in capillary endothelial cells is inhibited by the antiangiogenic factor 16-kDa N-terminal fragment of prolactin. *Proc. Natl. Acad. Sci. USA* 1995; 92:6374-6378.

[154] Rousseau S, Houle F, Landry J, Huot J. p38 MAP kinase activation by vascular endothelial growth factor mediates actin reorganization and cell migration in human endothelial cells. *Oncogene*. 1997;15(18):2169-77.

[155] Nakatsu MN, Sainson RC, Perez-del-Pulgar S, Aoto JN, Aitkenhead M, Taylor KL, Carpenter PM, Hughes CC. VEGF(121) and VEGF(165) regulate blood vessel diameter through vascular endothelial growth factor receptor 2 in an in vitro angiogenesis model. *Lab Invest* 2003;83:1873– 1885.

[156] Weiss RH, Ramirez A, Joo A. Short-term pravastatin mediates growth inhibition and apoptosis, independently of Ras, via the signaling proteins p27Kip1 and P13 kinase. *J Am Soc Nephrol.* 1999;10(9):1880-90.

[157] Negre-Aminou P, van Erck M, van Leeuwen RE, Collard JG, Cohen LH. Differential effect of simvastatin on various signal transduction intermediates in cultured human smooth muscle cells. *Biochem Pharmacol.* 2001;61(8):991-998.

[158] Sindermann JR, Schmidt A, Breithardt G, Buddecke E. Lovastatin controls signal transduction in vascular smooth muscle cells by modulating phosphorylation levels of mevalonate-independent pathways. *Basic Res Cardiol.* 2001;96(3):283-9.

[159] Takenaka M, Hirade K, Tanabe K, Akamatsu S, Dohi S, Matsuno H, Kozawa O. Simvastatin stimulates VEGF release via p44/p42 MAP kinase in vascular smooth muscle cells. *Biochem Biophys Res Commun.* 2003;301(1):198-203.

[160] Yamakawa T, Tanaka S, Kamei J, Kadonosono K, Okuda K. Pitavastatin inhibits vascular smooth muscle cell proliferation by inactivating extracellular signal-regulated kinases 1/2. *J Atheroscler Thromb.* 2003;10(1):37-42.

[161] Park JK, Muller DN, Mervaala EM, Dechend R, Fiebeler A, Schmidt F, Bieringer M, Schafer O, Lindschau C, Schneider W, Ganten D, Luft FC, Haller H. Cerivastatin prevents angiotensin II-induced renal injury independent of blood pressure- and cholesterol-lowering effects. *Kidney Int.* 2000;58:1420–1430.

[162] Takata R, Fukasawa S, Hara T, Nakajima H, Yamashina A, Yanase N, Mizuguchi J. Cerivastatin-induced apoptosis of human aortic smooth muscle cells through partial inhibition of basal activation of extracellular signal-regulated kinases. *Cardiovasc Pathol.* 2004;13(1):41-48.

[163] Horiuchi M, Cui T-X, Li Z, Li J-M, Nakagami H, Iwai M. Fluvastatin enhances the inhibitory effects of a selective angiotensin II type 1 receptor blocker, valsartan, on vascular neointimal formation. *Circulation* 2003;107:106-112.

[164] Sakamoto K, Murata T, Chuma H, Hori M, Ozaki H. Fluvastatin prevents vascular hyperplasia by inhibiting phenotype modulation and proliferation through extracellular signal-regulated kinase 1 and 2 and p38 mitogen-activated protein kinase inactivation in organ-cultured artery. *Arterioscler Thromb Vasc Biol.* 2005;25(2):327-333.

[165] Martinez-Gonzalez J, Escudero I, Badimon L. Simvastatin potenciates PGI(2) release induced by HDL in human VSMC: effect on Cox-2 up-regulation and MAPK signalling pathways activated by HDL. *Atherosclerosis.* 2004;174(2):305-313.

[166] Takahashi M, Ogata Y, Okazaki H, Takeuchi K, Kobayashi E, Ikeda U, Shimada K. Fluvastatin Enhances Apoptosis in Cytokine-Stimulated Vascular Smooth Muscle Cells. *J Cardiovasc Pharmacol.* 2002;39(2):310-317.

[167] Skaletz-Rorowski A, Muller JG, Kroke A, Waltenberger J, Pulawski E, Pinkernell K, Breithardt G. Lovastatin blocks basic fibroblast growth factor-induced mitogen-activated protein kinase signaling in coronary smooth muscle cells via phosphatase inhibition. Eur *J Cell Biol.* 2001;80(3):207-212.

[168] Kamiyama M, Utsunomiya K, Taniguchi K, Yokota T, Kurata H, Tajima N, Kondo K. Contribution of Rho A and Rho kinase to platelet-derived growth factor-BB-induced proliferation of vascular smooth muscle cells. *J Atheroscler Thromb.* 2003;10(2):117-123.

[169] Turner NA, O'Regan DJ, Ball SG, Porter KE. Simvastatin inhibits MMP-9 secretion from human saphenous vein smooth muscle cells by inhibiting the RhoA/ROCK pathway and reducing MMP-9 mRNA levels. *FASEB J.* 2005;19(7):804-806.

[170] Brandes RP, Beer S, Ha T, Busse R.Withdrawal of cerivastatin induces monocyte chemoattractant protein 1 and tissue factor expression in cultured vascular smooth muscle cells. *Arterioscler Thromb Vasc Biol.* 2003;23(10):1794-800.

[171] Tristano AG, Castejon AM, Castro A, Cubeddu LX. Effects of statin treatment and withdrawal on angiotensin II-induced phosphorylation of p38 MAPK and ERK1/2 in cultured vascular smooth muscle cells. *Biochem Biophys Res Commun.* 2007 Feb 2;353(1):11-7.

[172] Alessi DR, Cuenda A, Cohen P, Dudley DT, Saltiel AR. PD98059 is a specific inhibitor of the activation of mitogen-activated protein kinasein vitro and in vivo. *J Biol Chem* 1995;270:27489–27494.

[173] Favata MF, Horiuchi KY, Manos EJ, Daulerio AJ, Stradley DA, Feeser WS, Van Dyk DE, Pitts WJ, Earl RA, Hobbs F, Copeland RA, Magolda RL, Scherle PA, Trzaskos JM. Identification of a novel inhibitor of mitogen-activated protein kinase. *J Biol Chem.* 1998;273:18623–18632.

[174] Campbell M, Allen WE, Sawyer C, Vanhaesebroeck B, Trimble ER. Glucose-potentiated chemotaxis in human vascular smooth muscle is dependent on cross-talk between the PI3K and MAPK signaling pathways. *Circ Res.* 2004;95(4):380-388.

[175] Rouse J, Cohen P, Trigon S, Morange M, Alonso-Llamazares A, Zamanillo D, Hunt T, Nebreda AR. A novel kinase cascade triggered by stress and heat shock that stimulates MAPKAP kinase-2 and phosphorylation of the small heat shock proteins. *Cell* 1994;78:1027-37.

[176] Han J, Lee JD, Bibbs L, Ulevitch RJ. A MAP kinase targeted by endotoxin and hyperosmolarity in mammalian cells. *Science* 1994;265:808-11.

[177] Freshney NW, Rawlinson L, Guesdon F, Jones E, Cowley S, Hsuan J, Saklatvala J. Interleukin-1 activates a novel protein kinase cascade that results in the phosphorylation of Hsp27. *Cell* 1994;78:1039-1049.

[178] Lee JC, Laydon JT, McDonnell PC, Gallagher TF, Kumar S, Green D, McNulty D, Blumenthal MJ, Heys JR, Landvatter SW, Strickler JE, McLaughlin MM, Siemens IR, Fisher SM, Livi GP, White JR, Adams JL, Young PR. A protein kinase involved in the regulation of inflammatory cytokine biosynthesis. *Nature* 1994;372:739-746.

[179] Raingeaud J, Gupta S, Rogers JS, Dickens M, Han J, Ulevitch RJ, Davis RJ. Pro-inflammatory cytokines and environmental stress cause p38 mitogen-activated protein kinase activation by dual phosphorylation on tyrosine and threonine. *J Biol Chem* 1995;270:7420-6.

[180] Pietersma A, Tilly BC, Gaestel M, de Jong N, Lee JC, Koster JF, Sluiter W. p38 mitogen activated protein kinase regulates endothelial VCAM-1 expression at the post-transcriptional level. *Biochem Biophys Res Comm* 1997;230:44-48.

[181] Shalom-Barak T, Quach J, Lotz M. Interleukin-17-induced gene expression in articular chondrocytes is associated with activation of mitogen-activated protein kinases and NF-kappaB. *J Biol Chem* 1998;273:27467-73.

[182] Shapiro L, Puren AJ, Barton HA, Novick D, Peskind RL, Shenkar R, Gu Y, Su MS, Dinarello CA. Interleukin 18 stimulates HIV type 1 in monocytic cells. *Proc Natl Acad Sci U S A* 1998;95:12550-12555.

[183] Foltz IN, Lee JC, Young PR, Schrader JW. Hemopoietic growth factors with the exception of interleukin-4 activate the p38 mitogen- activated protein kinase pathway. *J Biol Chem* 1997;272:3296-3301.

[184] Kyriakis JM, Avruch J. Protein kinase cascades activated by stress and inflammatory cytokines. *BioEssays* 1996 ;18:567-577.

[185] Kusuhara M, Takahashi E, Peterson TE, Abe J, Ishida M, Han J, Ulevitch R, Berk BC. p38 Kinase is a negative regulator of angiotensin II signal transduction in vascular smooth muscle cells: Effects on Na1/H1 exchange and ERK 1/2. *Circ Res* 1998;83:824–831.

[186] Ushio-Fukai M, Griendling KK, Akers M, Lyons PR, Alexander RW. Temporal dispersion of activation of phospholipase C-beta 1 and –gamma isoforms by angiotensin II in vascular smooth muscle cells. Role of alphaq/11, alpha12, and beta gamma G protein subunits. *J Biol Chem* 1998;273:19772–19777.

[187] Wu G, Luo J, Rana JS, Laham R, Sellke FW, Li J. Involvement of COX-2 in VEGF-induced angiogenesis via P38 and JNK pathways in vascular endothelial cells. *Cardiovasc Res.* 2006;69(2):512-9.

[188] Lamalice L, Houle F, Jourdan G, Huot J. Phosphorylation of tyrosine 1214 on VEGFR2 is required for VEGF-induced activation of Cdc42 upstream of SAPK2/p38. *Oncogene* 2004;23:434– 45.

[189] Negre-Aminou P, van Leeuwen RE, van Thiel GC, van den IJssel P, de Jong WW, Quinlan RA, Cohen LH. Differential effect of simvastatin on activation of Rac1 vs. activation of the heat shock protein 27-mediated pathway upon oxidative stress, in human smooth muscle cells. *Biochem Pharmacol.* 2002;64(10):1483-91.

[190] Kyriakis JM, Banerjee P, Nikolakaki E, Dai T, Rubie EA, Ahmad MF, Avruch J, Woodgett JR. The stress-activated protein kinase subfamily of c-Jun kinases. *Nature* 1994;369:156-160.

[191] Pombo CP, Bonventre JV, Avruch J, Woodgett JR, Kyriakis JM, Force T. The stress-activated protein kinase (SAPKs) are major c-Jun amino-terminal kinases activated by ischemia and reperfusion. *J Biol Chem* 1994 ;269:26546-26551.

[192] Morooka H, Bonventre JV, Pombo CM, Kyriakis JM, Force T. Ischemia and reperfusion enhance ATF-2 and c-Jun binding to cAMP response elements and to an AP-1 binding site from the c-Jun promoter. *J Biol Chem* 1995;270:30084-30092.

[193] Xu Q, Liu Y, Gorospe M, Udelsman R, Holbrook J. Acute hypertension activate mitogen-activated protein kinases in arterial wall. *J Clin Invest* 1996;97:508-514.

[194] Zohn IE, Yu H, Li X, Cox AD, Earp HS. Angiotensin II stimulates calcium-dependent activation of c-Jun N-terminal kinase. *Mol Cell Biol* 1995;15:6160-6168.

[195] Shapiro PS, Evans JN, Davis RJ, Posada JA. The seven-transmembrane- spanning receptors for endothelin and thrombin cause proliferation of airway smooth muscle cells and activation of the extracellular regulated kinase and c-Jun NH2-terminal kinase groups of mitogen-activated protein kinases. *J Biol Chem* 1996;271:5750-5754.

[196] Araki S, Haneda M, Togawa M, Kikkawa R. Endothelin-1 activates c-Jun NH2-terminal kinase in mesangial cells. *Kidney Int* 1997;51:631-639.

[197] Force T, Pombo CM, Avruch JA, Bonventre JV, Kyriakis JM. Stress- activated protein kinases in cardiovascular disease. *Circ Res* 1996,78:947-953.

[198] Wen T, Scott S, Liu Y, Gonzales N, Nadler JL. Evidence that angiotensin II and lipoxygenase products activate c-Jun NH2-terminal kinase. *Circ Res* 1997;81:651–655.

[199] Schmitz U, Ishida T, Ishida M, Surapisitchat J, Hasham MI, Pelech S, Berk BC. Angiotensin II stimulates p21-activated kinase in vascular smooth muscle cells. Role in activation of JNK. *Circ Res* 1998;82:1272–1278.

[200] Kyaw M, Yoshizumi M, Tsuchiya K, Kagami S, Izawa Y, Fujita Y, Ali N, Kanematsu Y, Toida K, Ishimura K, Tamaki T. Src and Cas are essentially but differentially involved in angiotensin II-stimulated migration of vascular smooth muscle cells via extracellular signal-regulated kinase 1/2 and c-Jun NH2-terminal kinase activation. *Mol Pharmacol.* 2004;65(4):832-841.

[201] Libby P. Inflammation in atherosclerosis. *Nature* 2002; 420:868-874.

[202] Wick G, Knoflach M, Xu Q. Autoimmune and inflammatory mechanisms in atherosclerosis. *Annu. Rev. Immunol.* 2004;22:361-403.

[203] Badr K, Wainwright. Inflammation in the cardiovascular system: here, there and everywhere. *Curr opin. Pharmacol.* 2004;4:107-9.

[204] Doria A, Sherer Y, Meroni PL, Shoenfeld Y. Inflammation and accelerated atherosclerosis: Basic mechanisms. *Rheum. Dis. N. Am.* 2005;31:355-362.

[205] Paoletti R, Gotto AM, Hajjar DP. Inflammation in atherosclerosis and implications for the therapy. *Circulation* 2004; 109(suppl.III):III-20-III-26.

[206] Sherer Y, Shoenfeld Y. Mechanisms of disease: atherosclerosis in autoimmune diseases. *Nat Clin Pract Rheumatol.* 2006;2(2):99-106.

[207] Ross R. The pathogenesis of atherosclerosis: a perspective for the 1990s. *Nature* 1993; 362: 801-9.

[208] Aprahamian T, Rifkin I, Bonegio R, Hugel B, Freysinet JM, Sato K, Castellot JJ Jr., Walsh K. Impaired clearance of apoptotic cells promotes synergy between atherogeneis and autoimmune disease. *J Exp Med* 2004;199:1121–1131.

[209] Ward MM, Pyun E, Studenski S. Causes of death in systemic lupus erythematosus: long-term followup of an inception cohort. *Arthritis & Rheum* 1995;38:1492–9.

[210] Urowitz MB, Gladman DD, Abu-Shakra M, Farewell VT. Mortality studies in systemic lupus erythematosus: results from a single center. III. Improved survival over 24 years. *J Rheumatol* 1997;24:1061–5.

[211] Petri M, Perez-Gutthann S, Spence D, Hochberg MC. Risk factors for coronary artery disease in patients with systemic lupus erythematosus. *Am J Med* 1992;93:513–9.

[212] Rosner S, Ginzler Ginzler EM, Diamond HS, Weiner M, Schlesinger M, Fries JF, Wasner C, Medsger TA Jr, Ziegler G, Klippel JH, Hadler NM, Albert DA, Hess EV, Spencer-Green G, Grayzel A, Worth D, Hahn BH, Barnett EV. EM, Diamond HS, Weiner M, Schlesinger M, Fries JF, Wasner C, Medsger TA Jr, Ziegler G, Klippel JH, Hadler NM, Albert DA, Hess EV, Spencer-Green G, Grayzel A, Worth D, Hahn BH, Barnett EV. A multicenter study of outcome in systemic lupus erythematosus. II. Causes of death. *Arthritis Rheum* 1982;25:612–617

[213] Roman MJ, Shanker BA, Davis A Lockshin MD, Sammaritano L, Simantov R, Crow MK, Schwartz JE, Paget SA, Devereux RB, Salmon JE. Prevalence and correlates of accelerated atherosclerosis in systemic lupus erythematosus. *N Engl J Med* 2003;349:2399–406.

[214] Selzer F, Sutton-Tyrrell K, Fitzgerald SG, Pratt JE, Tracy RP, Kuller LH, Manzi S. Comparison of risk factors for vascular disease in the carotid artery and aorta in women with systemic lupus erythematosus. *Arthritis Rheum* 2004;50:151–9.

[215] Svenungsson E, Jensen-Urstad K, Heimburger M, Silveira A, Hamsten A, de Faire U, Witztum JL, Frostegard J. Risk factors for cardiovascular disease in systemic lupus erythematosus. *Circulation* 2001;104:1887–1893

[216] Doria A, Shoenfeld Y, Wu R, Gambari PF, Puato M, Ghirardello A, Gilburd B, Corbanese S, Patnaik M, Zampieri S, Peter JB, Favaretto E, Iaccarino L, Sherer Y, Todesco S, Pauletto P. Risk factors for subclinical atherosclerosis in a prospective cohort of patients with systemic lupus erythematosus. *Ann Rheum Dis* 2003;62:1071–7.

[217] De Inocencio J, Lovell DJ. Cardiac function in systemic lupus erythematosus. *J Rheumatol* 1994;21:2147–56.

[218] Petri M, Spence D, Bone LR, Hochberg MC. Coronary artery disease risk factors in the Johns Hopkins Lupus Cohort: prevalence, recognition by patients, and preventive practices. *Medicine (Baltimore)* 1992;71:291–302.

[219] Rahman P, Urowitz MB, Gladman DD, Bruce IN, Genest J Jr. Contribution of traditional risk factors to coronary artery disease in patients with systemic lupus erythematosus. *J Rheumatol* 1999;26:2363–8.

[220] Rahman P, Aguero S, Gladman DD, Hallett D, Urowitz MB. Vascular events in hypertensive patients with systemic lupus erythematosus. *Lupus* 2000;9:672–5.

[221] Bessant R, Hingorani A, Patel L, MacGregor A, Isenberg DA, Rahman A. Risk of coronary heart disease and stroke in a large British cohort of patients with systemic lupus erythematosus. *Rheumatology (Oxford)* 2004;43:924–9.

[222] Ridker PM, Hennekens CH, Buring JE, Rifai N. C-reactive protein and other markers of inflammation in the prediction of cardiovascular disease in women. *N Engl J Med* 2000;342:836–43.

[223] Toloza SM, Uribe AG, McGwin G Jr, Alarcon GS, Fessler BJ, Bastian HM, Vila LM, Wu R, Shoenfeld Y, Roseman JM, Reveille JD; LUMINA Study Group. Systemic

lupus erythematosus in a multiethnic US cohort (LUMINA). XXIII. Baseline predictors of vascular events. *Arthritis Rheum* 2004;50:3947–57.

[224] Bruce IN, Urowitz MB, Gladman DD, Ibanez D, Steiner G. Risk factors for coronary heart disease in women with systemic lupus erythematosus: the Toronto Risk Factor Study. *Arthritis Rheum* 2003;48:3159–67.

[225] Dessein PH, Stanwix AE, Joffe BI. Cardiovascular risk in rheumatoid arthritis versus osteoarthritis: acute phase response related decreased insulin sensitivity and high-density lipoprotein cholesterol as well as clustering of metabolic syndrome features in rheumatoid arthritis. *Arthritis Res* 2002;4:R5.

[226] Nawrocki AR, Scherer PE. The delicate balance between fat and muscle: adipokines in metabolic disease and musculoskeletal inflammation. *Curr Opin Pharmacol* 2004;4:281–9.

[227] Choi HK, Hernan MA, Seeger JD, Robins JM, Wolfe F. Methotrexate and mortality in patients with rheumatoid arthritis: a prospective study. *Lancet* 2002;359:1173–7.

[228] Booth AD, Jayne DR, Kharbanda RK, McEniery CM, Mackenzie IS, Brown J, Wilkinson IB. Infliximab improves endothelial dysfunction in systemic vasculitis: a model of vascular inflammation. *Circulation* 2004;109:1718–23.

[229] Hurlimann D, Forster A, Noll G, Enseleit F, Chenevard R, Distler O, Bechir M, Spieker LE, Neidhart M, Michel BA, Gay RE, Luscher TF, Gay S, Ruschitzka F. Anti-tumor necrosis factor-alpha treatment improves endothelial function in patients with rheumatoid arthritis. *Circulation* 2002;106:2184–7.

[230] Sattar N, Crompton P, Cherry L, Kane D, Lowe G, McInnes IB. Effects of tumor necrosis factor blockade on cardiovascular risk factors in psoriatic arthritis: a double-blind, placebo-controlled study. *Arthritis Rheum*. 2007;56(3):831-9.

[231] Ehrenstein M, Jury EC, Mauri C. Statins for atherosclerosis–as good as it gets? *N Engl J Med* 2005; 352: 73–75.

[232] Hillyard DZ, Cameron AJ, McDonald KJ, Thomson J, MacIntyre A, Shiels PG, Panarelli M, Jardine AG. Simvastatin inhibits lymphocyte function in normal subjects and patients with cardiovascular disease. *Atherosclerosis* 2004;175:305–313.

[233] Poloso NJ, Roche PA. Association of MHC class II-peptide complexes with plasma membrane microdomains. *Curr Opin Immunol* 2004; 16: 103–107.

[234] Emeson EE, Shen ML, Bell CG, Qureshi A. Inhibition of atherosclerosis in CD4 T-cell depleted and nude (nu/nu) C57BL/6 hyperlipidemic mice. *Am. J. Pathol.* 1996;149:675-85.

[235] Zhou X, Nicoletti A, Elhage R, Hansson GK. Transfer of CD4 + cells aggravates atherosclerosis in immunodeficient , apolipoprotein E knockout mice. *Circulation* 2000; 102:2919-22.

[236] Kabouridis PS, Janzen J, Magee AL, Ley SC. Cholesterol depletion disrupts lipid rafts and modulates the activity of multiple signalling pathways in T lymphocytes. *Eur J Immunol* 2000;30: 954–963.

[237] Xavier R, Brennan T, Li Q, McCormack C, Seed B. Membrane compartmentation is required for efficient T cell activation. *Immunity* 1998;8:723–732.

[238] Goldman F, Hohl RJ, Crabtree J, Lewis-Tibesar K, Koretzky G. Lovastatin inhibits T-cell antigen receptor signaling independent of its effects on ras. *Blood* 1996;88:4611–4619.

[239] Hillyard DZ, Jardine AG, McDonald KJ, Cameron AJ. Fluvastatin inhibits raft dependent Fcgamma receptor signalling in human monocytes. *Atherosclerosis* 2004;172:219–228.

[240] Jury EC, Ehrenstein MR. Statins: immunomodulators for autoimmune rheumatic disease? *Lupus.* 2005;14(3):192-6.

[241] Bacon PA, Raza K, Banks MJ, Townend J,Kitas GD. The role of endothelial cell dysfunction in the cardiovascular mortality of RA. *Int Rev Immunol* 2002;21:1–17.

[242] Riboldi P, Gerosa M, Meroni PL. Statins and autoimmune diseases. *Lupus* 2005;14:765-768.

[243] Abud-Mendoza C, de la Fuente H, Cuevas-Orta E, Baranda L, Cruz-Rizo J, Gonzalez-Amaro R. Therapy with statins in patients with refractory rheumatic diseases: a preliminary study. *Lupus* 2003;12:607–11.

[244] Lawman S, Mauri C, Jury EC, Cook HT, Ehrenstein MR. Atorvastatin inhibits autoreactive B cell activation and delays lupus development in New Zealand black/white F1 mice. *J Immunol* 2004;173:7641–6.

[245] McCarey DW, McInnes IB, Madhok R, Hampson R, Scherbakov O, Ford I, Capell HA, Sattar N.. Trial of atorvastatin in rheumatoid arthritis (TARA): double-blind, randomised placebo-controlled trial. *Lancet* 2004;363:2015–21.

[246] Kanda H, Hamasaki K, Kubo K, Tateishi S, Yonezumi A, Kanda Y, Yamamoto K, Mimura T. Antiinflammatory effect of simvastatin in patients with rheumatoid arthritis. *J Rheumatol* 2002; 29:2024–6.

[247] ten Cate R, Nibbering PH, Bredius RG. Therapy-refractory systemic juvenile idiopathic arthritis successfully treated with statins. *Rheumatology (Oxford)* 2004;43:934–5.

[248] Diomede L, Albani D, Sottocorno M, Donati MB, Bianchi M, Fruscella P, Salmona M. In vivo anti-inflammatory effect of statins is mediated by nonsterol mevalonate products. *Arterioscler Thromb Vasc Biol* 2001;21:1327–32.

[249] Kimura M, Kurose I, Russell J, Granger DN. Effects of fluvastatin on leukocyte-endothelial cell adhesion in hypercholesterolemic rats. *Arterioscler Thromb Vasc Biol* 1997;17:1521–6.

[250] Stalker TJ, Lefer AM, Scalia R. A new HMG-CoA reductase inhibitor, rosuvastatin, exerts anti-inflammatory effects on the microvascular endothelium: the role of mevalonic acid. *Br J Pharmacol* 2001;133:406–12.

[251] Sparrow CP, Burton CA, Hernandez M, Mundt S, Hassing H, Patel S, Rosa R, Hermanowski-Vosatka A, Wang PR, Zhang D, Peterson L, Detmers PA, Chao YS, Wright SD. Simvastatin has anti-inflammatory and antiatherosclerotic activities independent of plasma cholesterol lowering. *Arterioscler Thromb Vasc Biol* 2001;21:115–21.

[252] Aktas O, Waiczies S, Smorodchenko A, Dorr J, Seeger B, Prozorovski T, Sallach S, Endres M, Brocke S, Nitsch R, Zipp F. Treatment of relapsing paralysis in experimental encephalomyelitis by targeting Th1 cells through atorvastatin. *J Exp Med* 2003;197:725–33.

[253] Nath N, Giri S, Prasad R, Singh AK, Singh I. Potential targets of 3-hydroxy-3-methylglutaryl coenzyme A reductase inhibitor for multiple sclerosis therapy. *J Immunol* 2004;172:1273–86.

[254] Greenwood J,Walters CE, Pryce G, Kanuga N, Beraud E, Baker D, Adamson P. Lovastatin inhibits brain endothelial cell Rho-mediated lymphocyte migration and attenuates experimental autoimmune encephalomyelitis. *FASEB J* 2003;17:905–7.

[255] Stanislaus R, Singh AK, Singh I. Lovastatin treatment decreases mononuclear cell infiltration into the CNS of Lewis rats with experimental allergic encephalomyelitis. *J Neurosci Res* 2001;66:155–162

[256] Veillard NR, Braunersreuther V, Arnaud C, Burger F, Pelli G, Steffens S, Mach F. Simvastatin modulates chemokine and chemokine receptor expression by geranylgeranyl isoprenoid pathway in human endothelial cells and macrophages. *Atherosclerosis* 2006;188:51–8.

[257] Ghittoni R, Napolitani G, Benati D, Ulivieri C, Patrussi L, Laghi Pasini F, Lanzavecchia A, Baldari CT. Simvastatin inhibits the MHC class II pathway of antigen presentation by impairing Ras superfamily GTPases. *Eur J Immunol.* 2006;36(11):2885-93.

[258] Rudich SM, Mongini PK, Perez RV, Katznelson S. HMG-CoA reductase inhibitors pravastatin and simvastatin inhibit human B-lymphocyte activation. *Transplant Proc.* 1998;30(4):992-5.

[259] Samson KT, Minoguchi K, Tanaka A, Oda N, Yokoe T, Okada S, Yamamoto Y, Watanabe Y, Yamamoto M, Ohta S, Adachi M. Effect of fluvastatin on apoptosis in human CD4+ T cells. *Cell Immunol* 2005;235:136–44.

[260] Kaneta S, Satoh K, Kano S, Kanda M, Ichihara K. All hydrophobic HMG-CoA reductase inhibitors induce apoptotic death in rat pulmonary vein endothelial cells. *Atherosclerosis* 2003;170:237–43.

[261] Li X, Liu L, Tupper JC, Bannerman DD, Winn RK, Sebti SM, Hamilton AD, Harlan JM. Inhibition of protein geranylgeranylation and RhoA/RhoA kinase pathway induces apoptosis in human endothelial cells. *J Biol Chem* 2002;277:15309–16.

[262] Blanco-Colio LM, Villa A, Ortego M, Hernandez-Presa MA, Pascual A, Plaza JJ, Eqido J. 3-Hydroxy-3-methyl-glutaryl coenzyme A reductase inhibitors, atrovastatin and simvastatin, induce apoptosis of vascular smooth muscle cells by downregulation of Bcl-2 expression and Rho prenylation. *Atherosclerosis* 2002;161:17–26.

[263] Guijarro C, Blanco-Colio LM, Ortego M, Alonso C, Ortiz A, Plaza JJ, Diaz C, Hernandez G, Eqido J. 3-Hydroxy-3-methylglutaryl coenzyme A reductase and isoprenylation inhibitors induce apoptosis of vascular smooth muscle cells in culture. *Circ Res* 1998;83: 490–500.

[264] Hillyard DZ, Cameron AJ, McIntyre AH, Hadden MH, Marshall HE, Johnston N, Jardine AG. Inhibition of proliferation and signalling mechanisms in human lymphocytes by fluvastatin. *Clin Exp Pharmacol Physiol* 2002;29:673–8.

[265] Chapman-Shimshoni D, Yuklea M, Radnay J, Shapiro H, Lishner M. Simvastatin induces apoptosis of B-CLL cells by activation of mitochondrial caspase 9. *Exp Hematol.* 2003;31(9):779-83.

[266] Rezaie-Majd A, Prager GW, Bucek RA, Schernthaner GH, Maca T, Kress HG, Valent P, Binder BR, Minar E, Baghestanian M. Simvastatin reduces the expression of adhesion molecules in circulating monocytes from hypercholesterolemic patients. *Arterioscler Thromb Vasc Biol* 2003;23:397–403

[267] Grip O, Janciauskiene S, Lindgren S. Atorvastatin activates PPAR-gamma and attenuates the inflammatory response in human monocytes. *Inflamm Res* 2002;51:58–62.

[268] Frigerio S, Gelati M, Boncoraglio G, Ciusani E, Croci D, de Curtis M, Parati E, Salmaggi A. Pravastatin in vivo reduces mononuclear cell migration through endothelial monolayers. *Neurol Sci*. 2006;27(4):261-5.

[269] Tristano AG, Fuller K. Immunomodulatory effects of statins and autoimmune rheumatic diseases: novel intracellular mechanism involved. *Int Immunopharmacol*. 2006;6(12):1833-46.

[270] Mok CC, Lau CS. Pathogenesis of systemic lupus erythematosus. *J Clin Pathol* 2003;56:481-490.

INDEX

#

A

E

F

G

H

I

M

N

O

P

Q

R

S

T

U

V

W

X

Y

Z